Conference Proceedings of the Society for Experimental Mechanics Series

Series Editor

Kristin B. Zimmerman, Ph.D.
Society for Experimental Mechanics, Inc.,
Bethel, CT, USA

The Conference Proceedings of the Society for Experimental Mechanics Series presents early findings and case studies from a wide range of fundamental and applied work across the broad range of fields that comprise Experimental Mechanics. Series volumes follow the principle tracks or focus topics featured in each of the Society's two annual conferences: IMAC, A Conference and Exposition on Structural Dynamics, and the Society's Annual Conference & Exposition and will address critical areas of interest to researchers and design engineers working in all areas of Structural Dynamics, Solid Mechanics and Materials Research.

More information about this series at http://www.springer.com/series/8922

Jacob Notbohm
Nikhil Karanjgaokar • Christian Franck
Frank W. DelRio
Editors

Mechanics of Biological Systems and Materials & Micro-and Nanomechanics & Research Applications

Proceedings of the 2020 Annual Conference on Experimental and Applied Mechanics

Springer

Editors
Jacob Notbohm
533 Engineering Research Building
University of Wisconsin-Madison
Madison, WI, USA

Nikhil Karanjgaokar
242 Higgins Lab
Worchester Polytech Institute
Worcester, MA, USA

Christian Franck
Brown University
Providence, RI, USA

Frank W. DelRio
Sandia National Laboratories
Albuquerque, MN, USA

ISSN 2191-5644 ISSN 2191-5652 (electronic)
Conference Proceedings of the Society for Experimental Mechanics Series
ISBN 978-3-030-59764-1 ISBN 978-3-030-59765-8 (eBook)
https://doi.org/10.1007/978-3-030-59765-8

This Springer imprint is published by the registered company Springer Nature Switzerland AG
The registered company address is: Gewerbestrasse 11, 6330 Cham, Switzerland

Preface

Mechanics of Biological Systems and Materials & Micro- and Nanomechanics & Research Applications represents one of seven volumes of technical papers to be presented at the 2020 SEM Annual Conference & Exposition on Experimental and Applied Mechanics organized by the Society for Experimental Mechanics scheduled to be held in Orlando, FL, September 14–17, 2020. The complete Proceedings also includes volumes on *Dynamic Behavior of Materials; Challenges in Mechanics of Time-Dependent Materials; Fracture, Fatigue, Failure and Damage Evolution; Advancement of Optical Methods & Digital Image Correlation in Experimental Mechanics; Mechanics of Composite, Hybrid & Multifunctional Materials; and Thermomechanics & Infrared Imaging, Inverse Problem Methodologies and Mechanics of Additive & Advanced Manufactured Materials.*

Each collection presents early findings from experimental and computational investigations on an important area within experimental mechanics, the mechanics of biological systems and materials, micro- and nanomechanics, and other experimental and applied mechanics such as research in progress.

The biological systems and materials segment of this volume summarizes the exchange of ideas and information among scientists and engineers involved in the research and analysis of how mechanical loads interact with the structure, properties, and function of living organisms and their tissues. The scope includes experimental, imaging, numerical, and mathematical techniques and tools spanning various lengths and timescales. Establishing this symposium at the Annual Meeting of the Society for Experimental Mechanics provides a venue where state-of-the-art experimental methods can be leveraged in the study of biological and bioinspired materials, traumatic brain injury, cell mechanics, and biomechanics in general. A major goal of the symposium was for participants to collaborate in the asking of fundamental questions and the development of new techniques to address bioinspired problems in society, human health, and the natural world. The 2020 Symposium is the 10th International Symposium on the Mechanics of Biological Systems and Materials. The organizers would like to thank all the speakers and staff at SEM for enabling a successful program.

The micro- and nanomechanics segment of this volume focuses on specialized scientific areas that involve miniaturizing conventional scale components and systems to take advantage of reduced size and weight and/or enhanced performance or novel functionality. These fields also encompass the application of principles ranging from the micron scale down to individual atoms. Sometimes these principles borrow from conventional scale laws but often involve new physical and/or chemical phenomena that require new behavioral laws and impart new properties to exploit. Studying how mechanical loads interact with components of these scales is important in developing new applications, as well as assessing their reliability and functionality. Establishing this symposium at the Annual Meeting of the Society for Experimental Mechanics provides a venue where state-of-the-art experimental methods can be leveraged in these endeavors.

The 2020 Symposium is the 21st in the series and addresses pertinent issues relating to design, analysis, fabrication, testing, optimization, and applications of micro- and nanomechanics, especially as these issues relate to experimental mechanics of microscale and nanoscale structures.

It is with deep gratitude that we thank the Organizing Committee, Session Chairs, Authors and Keynote Speakers, Participants, and SEM Staff for making the 21st *International Symposium on Micro- and Nanomechanics (ISMAN)* and the *10th International Symposium on the Mechanics of Biological Systems and Materials* a valuable and unforgettable experience.

Madison, WI	Jacob Notbohm
Worcester, MA	Nikhil Karanjgaokar
Madison, WI	Christian Franck
Boulder, CO	Frank W. DelRio

Contents

Determination of Texture Properties of White Long Turnip

1

Deepak Kumar Mishra, Quoc Thien Pham, and Nai-Shang Liou

Abstract

The purpose of this study was to determine the texture and mechanical properties of white long turnip (also known as a radish) by using a double-bite compression test. The texture parameters were quantified by using the double-bite compression test. Double bite tests were applied to cylindrical specimens of turnip flesh. Samples were taken from a well-defined portion (upper portion, a middle portion, lower portion). The samples were prepared in the longitudinal and radial directions relative to the long axis of a turnip. The test parameters were determined through the evaluation of the force-time curves of experiments. The test was repeatedly conducted using a custom made testing machine. Texture properties such as hardness, cohesiveness, springiness, and cohesiveness were obtained. The results show that hardness, fracturability, springiness and chewiness of specimens made in longitudinal direction are higher than those of specimens made in the radial direction. However, cohesiveness and resilience of specimens made in longitudinal direction are lower than those of specimens made in the radial direction. The result obtained in this study is useful in pickle process of white long turnip to improve the quality.

Keywords

Double-bite test , Texture profile analysis , Biomechanics , Mechanical test , Fracturability

1.1 Introduction

The turnip is a biennial plant that is most commonly grown as an annual plant. Turnip scientific name is either Brassica Rapa or Brassica campestris. Turnips are nearly perfectly round with white flesh and thin rough leaves covered with prickly hair. The processing of vegetables is an important portion of the food economy in Taiwan. The quality of the supplied raw turnip is important for the food processing factory because it directly affects the quality of processed products. Turnips are primarily used as an ingredient in soups and stews often consumed during the winter, but they also have other potential uses. Turnips can be served as a mashed dish, baked, fried, boiled or used for winemaking. Turnips can be stored 4–5 days in the refrigerator and 2–3 days outside the refrigerator. After a few days of storage turnip getting soft and its taste is less likely compared to freshen ones. In order to properly process raw turnip, it is necessary to precisely determine mechanical or texture properties. From the point of view of consumers, texture properties are also an important attribute for estimating the quality of food. Textural properties can be calculated using a texture profile analyzer (TPA), which helps to calculate desirable characteristics in quality control and product development. The TPA procedure provides double-bite compression on food specimens at a compression speed and distance to obtain certain mechanical properties of food which can be interpreted

D. K. Mishra · Q. T. Pham · N.-S. Liou (✉)
Department of Mechanical Engineering, Southern Taiwan University of Science and Technology, Tainan, Taiwan, R.O.C.
e-mail: nliou@stust.edu.tw

J. Notbohm et al. (eds.), *Mechanics of Biological Systems and Materials & Micro-and Nanomechanics & Research Applications*, Conference Proceedings of the Society for Experimental Mechanics Series,
https://doi.org/10.1007/978-3-030-59765-8_1

as hardness, cohesiveness, chewiness, fracturability, springiness, gumminess and resilience [1, 2]. Despite numerous scientific studies in this field, the available data on the physical properties of the raw materials of turnip are still of limited. Thus, in this study, the texture properties of raw turnip were investigated.

1.2 Materials and Methods

The White long turnips were used for this research (Fig. 1.1). The turnips for test were collected fresh from a local market in Tainan, Taiwan, and stored at 17 °C. Cylindrical specimens (10 × 10 mm) were taken from the upper, middle and lower portion of the long turnip in a longitudinal and radial direction. Before tests, specimens were stored in the Petri platter with little water. The tests were carried out within 15 min of the preparation of the specimens. The TPA test was carried out using a custom made TPA with a 50 kg load cell. The upper crosshead was mounted on a linear motion stage and powered by the stepper motor.

The experimental setup is shown in Fig. 1.2. All tests were carried out within 2 days of their purchase. The samples were cut in a strictly defined manner due to the diverse and different structure of the white turnip. Figure 1.3 demonstrates the appearance of the turnip structure after cutting in radial and longitudinal directions. The turnip was assumed to be symmetrical with respect to the longitudinal axis.

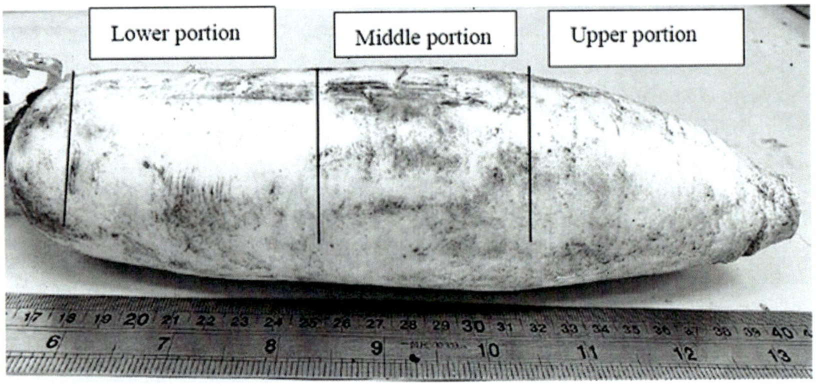

Fig. 1.1 Cylindrical specimen taken from upper, middle and lower portion of turnip

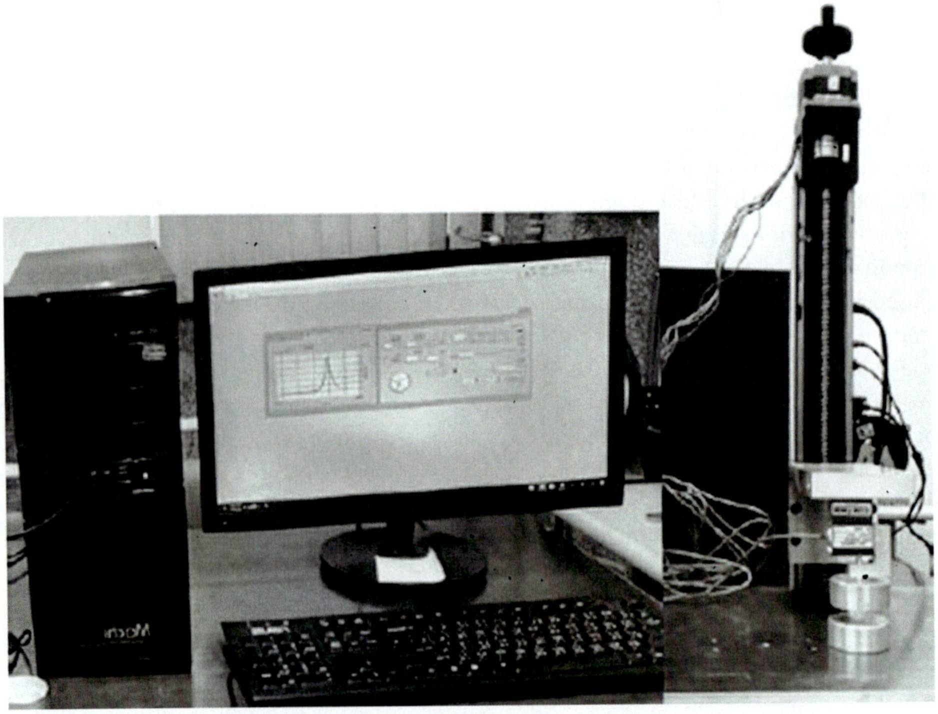

Fig. 1.2 Experimental setup

Fig. 1.3 Specimens from white long turnip: (**a**) Longitudinal and (**b**) Radial directions

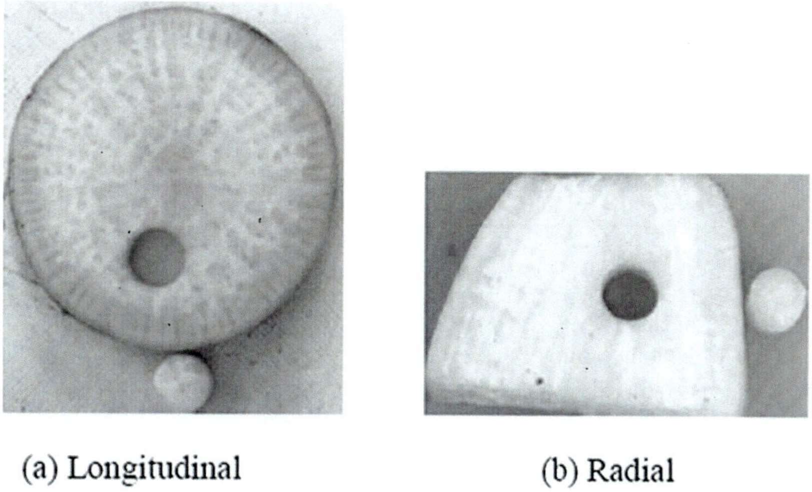

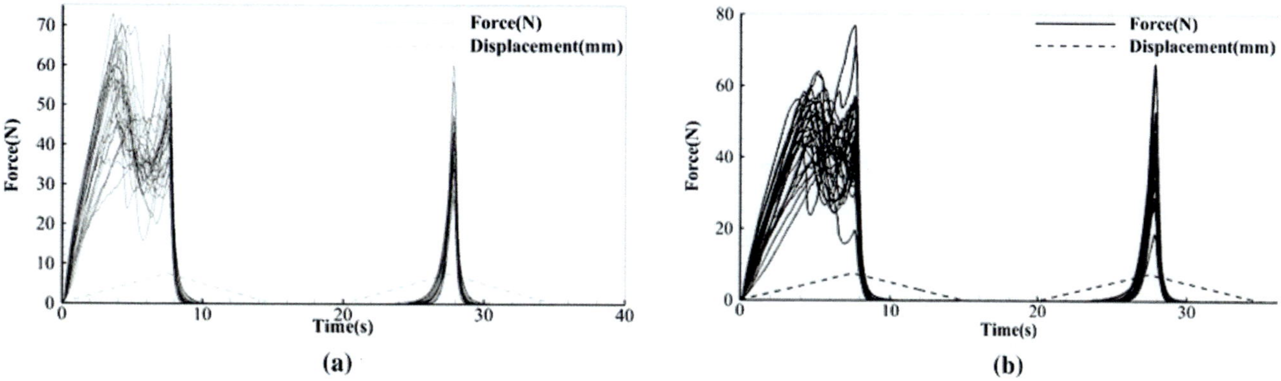

Fig. 1.4 Force-time curves of longitudinal samples made in (**a**) longitudinal and (**b**) radial directions

The parameters of white turnip specimens collected from TPA are hardness, cohesiveness, fracturability, springiness, chewiness, and resilience. The hardness is the peak intensity at which the first compression occurs. Fracturability is the strength of the first point. Cohesiveness is the area of work during the second compression divided by the area of work during the first compression. Springiness is expressed by the distance of the detected height during the second compression divided by the original compression distance. The chewiness is calculated by multiplying the hardness, cohesiveness and springiness values. Resilience is calculated by dividing the first compression upstream energy by the downstream energy of the first compression [2]. The white long turnip samples were compressed twice to 75% of their original height at a displacement rate of 0.5 mm/s. The s between the two compression periods is 5 s.

1.3 Results and Discussion

The force-time curves of the TPA test were shown in Fig. 1.4. The mean values of TPA test parameters of white long turnip from two directions are listed in Table 1.1 respectively. For both directions, there is a difference in values of TPA parameters of longitudinal and radial. The results show that hardness, fracturability, springiness and chewiness of specimens made in longitudinal direction are higher than those of specimens made in the radial direction. However, cohesiveness and resilience of specimens made in longitudinal direction are lower than those of specimens made in the radial direction.

TPA imitates the way food products are exposed to an instrument in mouth and TPA's concept of texture is a sensory factor. Fundamental mechanical test parameters such as elastic modulus, ultimate strength, and fracture strain can also be associated with food sensory attributes. The use of cylindrical pear flesh specimens for TPA processing, the strength at the fracture and ultimate strength can be calculated by diving the fracturability and hardness with the cross section area respectively. The elastic modulus E can be determined from the linear component of stress-strain relations deduced from the first step of TPA studies of force-displacement relationships, and the computed results showed.

Table 1.1 TPA test parameters of white long Turnip for 7.5 mm compression displacement

	Ha	Fr	Co	Sp	Ch	Re
Longitudinal	57.83	56.69	0.115	0.395	2.584	0.091
Radial	52.93	51.962	0.132	0.349	2.424	0.094

Ha Hardness, *Fr* fracturability, *Co* cohesiveness, *Sp* springiness, *Ch* chewiness, *Re* resilience

1.4 Conclusion

The TPA testing and analysis of white long turnip showed different texture and mechanical properties for two different directions. The white long turnip flesh samples have higher hardness, factorability and chewiness in longitudinal direction specimen while the radial has the lower values. The fiber orientation in the sample did not significantly affect the elasticity modulus value. White long turnip cohesion is higher when samples were compressed in radial direction. The result obtained in this study is useful in pickle process of white long turnip to improve the quality.

References

1. Alvarez, M.D., Canet, W., Lopez, M.E.: Influence of deformation rate and degree of compression on textural parameters of potato and apple tissues in texture profile analysis. Eur. Food Res. Technol. **215**, 13–20 (2002)
2. Singh, V., Guizani, N., Al-Alawi, A., Claereboudt, M., Rahman, M.S.: Instrumental texture profile analysis (TPA) of date fruits as a function of its physico-chemical properties. Ind. Crop. Prod. **50**, 866–873 (2013)

2 Impact Testing of a Commercial Poly-Lactic Acid

George N. Khawly, Nicolas R. Fabbri,
Amber J. W. McClung, and Juan D. Ocampo

Abstract

As the fabrication of prototypes continues to become ever more accessible to wider audiences, and advances in additive manufacturing techniques such as 3D printing progresses, it is imperative to understand the properties of materials adopted. One specific material: Polylactic acid (PLA), is taking the spotlight for its cost efficiency, availability and biodegradable nature. The lack of characterization and variability in mechanical properties of 3D printed thermoplastics proves a hindrance in the application of this material. To address this issue, the current research aims to characterize PLA utilizing standard ISO 179-1:2010 while varying printing parameters such as infill density, and printing speed. This investigation focuses on how the afore mentioned properties affect the resulting impact energy of the polymer. Standard manufactured filament acquired from K-camel was extruded through a fused deposition 3D printer (Prusa Geetech I3) and tested on a CEAST 9050 impact testing machine. The result also showed a decrease in impact energy as infill density decreases at most of the print speeds tested as expected. The impact energy however did not show a clear pattern as print speed increases. Future work includes exploring the effect of the material cooling rate to identify if it explains the current patterns in mechanical performance presented here and in the published literature at large.

Keywords

Polymers , Polylactic acid , 3D printing , Filament fused deposition , Impact testing

2.1 Introduction and Background

In recent years, 3D printing has become one of the most promising areas in production technologies. Although this spotlight has paved the way for a large amount of investment, regulations are still relatively preliminary, and research still focuses on areas that have been defined in other manufacturing methods for a long time. Comprehensive characterization plans are yet to be developed to understand the array of material properties that can be produced during fused deposition printing.

While research investigating 3D printing and impact properties are not readily found in the published literature, efforts can be easily found exploring other various aspects of 3D printing. Torres et al. [1] researched the material properties of PLA through monotonic tension of dogbone samples and through fracture testing. They showed that most of the mechanical properties in both types of testing depend on several print parameters, with print density ranking as the top influencer on most of the measured properties. Jerez-Mesa et al. [2] explored PLA through fatigue studies. They also explored variations in printer parameters and settings. Gordeev et al. [3] focused on the permeability of the prints with regards to the different parameters set. This permeability work did show that changing the print settings does changes the quality of the samples. They varied filament feed rate, wall geometry and G-code-defined wall structure to see their effects of the samples permeability to air.

G. N. Khawly · N. R. Fabbri · A. J. W. McClung (✉) · J. D. Ocampo
Department of Engineering, School of Science, Engineering, and Technology, St. Mary's University, San Antonio, TX, USA
e-mail: amcclung@stmarytx.edu

J. Notbohm et al. (eds.), *Mechanics of Biological Systems and Materials & Micro-and Nanomechanics & Research Applications*, Conference Proceedings of the Society for Experimental Mechanics Series,
https://doi.org/10.1007/978-3-030-59765-8_2

They interestingly submerged 3D printed parts in water and pumped air through them. Imperfections existed that could not be seen by the naked eye but could have easily been solved by the rate at which filament was extruded. Another article by Cantrel et al. details specifically how important print orientation is to the overall integrity of the printed product [4]. They noted that a rotation of 45 degrees could mean the difference between almost double the integrity.

The thermal properties of PLA have been studies as well. Matos et al. [5] evaluated the thermal properties of various commercial filaments through TGA and DSC. They also explored the rheology of the various samples. No mechanical characterization was included in this work. However, the work could help explain mechanical property findings in the future if correlated with the heat transfer processes occurring during the 3D printing process.

The authors predict that the heat transfer process during printing drives the changes in material properties. A presentation from the authors at SEM 2018 studied the effects of temperature during the printing process as a printing parameter that is not directly controlled through the print software [6]. The yield stress and maximum stress did increase with increasing print temperature. The maximum strain decreased by a little more than half when printed at all temperatures above 25 °C. In the long term, the changes in the heat transfer process caused by varying print parameters could be used as the governing equations to predict changes in material mechanical properties.

The research objective of this current paper is to investigate the influence of print speed and infill density during printing on the mechanical behavior of the 3D printed polymers. Experimental results generated here give insight into the changes in the inelastic behavior of the polymer with changes in these two print settings. This insight will provide a foundation for understanding material property sensitivity to these setting values.

2.2 Equipment and Materials

The material printed is a 1.75 mm diameter PLA (Polylactic Acid) filament manufactured by K-Camel. The samples required for testing were all printed using a Geeetech Prusa I3 model. This model printer uses Fused Filament Fabrication. Layer resolution was set at a first layer height of 0.3 mm with all other layers set at 0.2 mm height. The feed filament diameter is 1.75 mm, and the diameter is 0.3 mm. The experimental samples were printed at varying speeds of 40, 50, 60 and 70 mm/s. Infill percentages for each speed were set at 0, 25, 50, 75, and 100%, filling in the complete range of the printer's capabilities. Three samples were tested for each combination of print speed and infill percentage.

Impact testing was performed on an Instron CEAST 9050 Pendulum Impact System with a 50 J Charpy impact hammer. The experimental procedure was conducted following ISO 179-1 for geometry and orientation. ISO 179-1 details the standard procedures for Charpy Impact Testing of plastic materials [7]; Subsection 1.5 states that notched samples are not normally used with rigid cellular materials. The 3D printing process creates a material with voids similar to a cellular structure, therefore notches are not used in this study. Specimen type 1 is used which dictates a geometry of 80 mm in length, 10 mm wide, and 4 mm thick. Two different orientations are generally used in Charpy Impact testing. The two styles are edgewise, and flatwise. Flatwise impact occurs along the width of the specimen (10 mm), while edgewise impact occurs along the height of the specimen (4 mm). In the current study, flatwise impact was used due to interest in concentrating energy in the surface (most densely printed portion) of all of the samples.

2.3 Data and Analysis

Several preliminary tests have been run so far, including the control group (printed at 40 mm/s) with the isolated variable being infill percentage and another baseline group of infill percentage with varying print speed. For the infill percentage baseline shown in Fig 2.1a, the behavior followed the expected trend, as the printer printed the samples with less and less material (lower infill percentage), the energy each sample was able to absorb decreased. For the samples printed at the varying rates (Fig. 2.1), the trends followed those found in the 40 mm/s; as the infill increases, the energy absorbed also increases.

As for the comparison tests of energy versus speed, the results (Fig. 2.2) are rather interesting. For the high infill regime of 100% infill, the absorbed energy drops as the print speed increases. For the infills of 50% and less, it looks like the absorbed energy stays fairly constant regardless of print speed. And at the 75% infill, the material seems to be in a transition between the two low infill and hi infill behaviors as the results jump up and down as print speed increases.

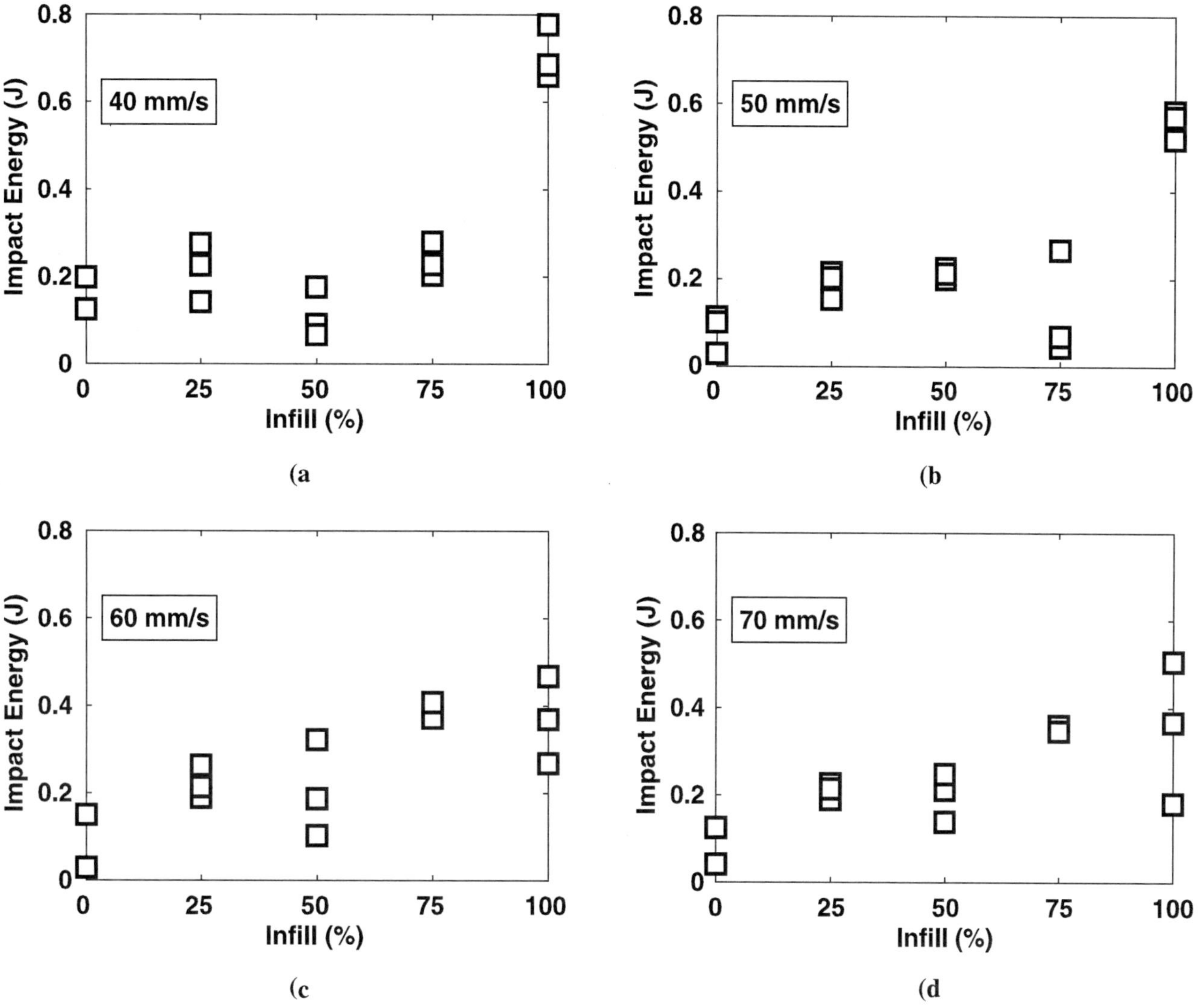

Fig. 2.1 Impact energy versus percentages of infill at a print speeds of (**a**) 40 mm/s, (**b**) 50 mm/s, (**c**) 60 mm/s, (**d**) 70 mm/s

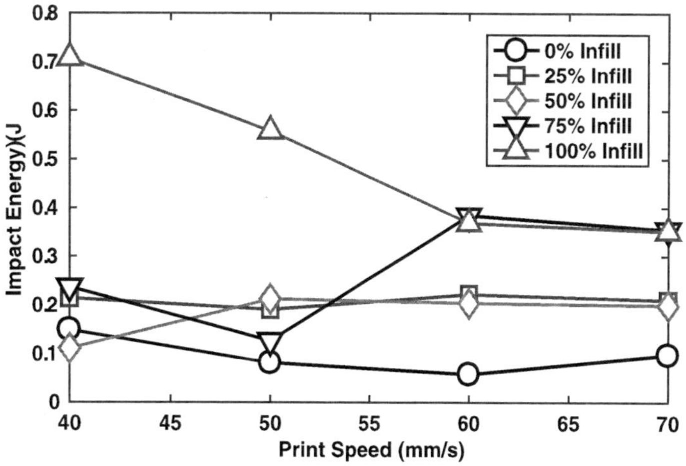

Fig. 2.2 Average absorbed energy versus speed of print head

2.4 Ongoing Work

Ongoing work includes probabilistic analysis to determine the statistical correlation between the print speed, infill percentage, and the impact energy. The goal is to create a 3D surface that can predict what impact energy the material will exhibit with a given print speed and infill setting.

Another ongoing branch of this research is exploring the addition of copper powder to PLA filaments. The goal is to explore how much the copper changes the heat transfer properties of the filament and whether the addition of the copper measurably changes the mechanical properties of the resulting prints.

2.5 Conclusion

Samples of PLA were 3D printed with infill varying infill settings (from 0% to 100%) and varying print speeds (from 40 mm to 70 mm). The samples were Charpy impact samples following standard ISO 179-1 geometry. Charpy impact tests were conducted on the samples in a flatwise impact orientation. For the infill percentage, the behavior followed the expected trend, as the printer printed the samples with more and more material (higher infill percentage), the energy each sample was able to absorb increased. However, as the print speed is increased, the material exhibits a drop in impact energy as the print speed increases for the highest infill group, but no discernable influence on impact energy at the low infill group. The patterns are initial results in an ongoing study.

Acknowledgements Research reported in this publication was supported by the National Institutes of Health, the U.S. Department of Education, and St. Mary's University.

The support from the National Institute of General Medical Sciences of the National Institutes of Health Department is under Award number 5T34GM008073-35 to St. Mary's University. The content is solely the responsibility of the authors and does not necessarily represent the official views of the National Institutes of Health.

This project was supported in part by a US Department of Education Ronald E. McNair Post Baccalaureate Achievement Grant to St. Mary's University of San Antonio, TX (P217A170332).

References

1. Torres, J., Cole, M., Owji, A., DeMastry, Z., Gordon, A.P.: An approach for mechanical property optimization of fused deposition modeling with polylacic acid via design of experiments. Rapid Prototyp. **22**(2), 387–404 (2016)
2. Jerez-Mesa, R., Travieso-Rodirguez, J.A., Lluma-Fuentes, J., Gomez-Gras, G., Puig, D.: Fatigue lifespan study of PLA parts obtained by additive manufacturing. Proc. Manuf. **13**, 872–879 (2017)
3. Gordeev, E.G., Galushko, A.S., Ananikov, V.P.: Improvement of quality of 3D printed objects by elimination of microscopic structural defects in fused deposition modeling. PLoS One. **13**(6), 1–19 (2018)
4. Cantrell, J.T., Rohde, S., Damiani, D., Gurnani, R., DiSandro, L.: Experimental characterization of the mechanical properties of 3-D printed ABS and polycarbonate parts. Rapid Prototyp. J. **23**(4), 811–824 (2017)
5. Matos, B.D.M., Rocha, V., da Silva, E.J., Moro, F.H., Bottene, A.C., al, e.: Evaluation of commercially available polylactic acid (PLA) filamentsfor 3D printing applications. J. Therm. Anal. Calorim. **137**(2), 555–562 (2019)
6. Torres, J., Onwuzurike, O., McClung, A.J., Ocampo, J.D.: Influence of an extreme on the tensile mechanical properties of a 3D printed thermoplastic polymer, extended abstract, in Society for Experimental Mechanics, Greenville SC, (2018)
7. International Standards Organization. ISO 179-1: Plastics: determination of charpy impact properties. Part 1 non-instrumental impact test. (2010)

Explosive Bubble Interaction with an Adjacent Underwater Structure

3

Carlos Javier, Michael Galuska, Michael Papa, James LeBlanc, Helio Matos, and Arun Shukla

Abstract

The interaction between structures and bubbles generated by an underwater explosive (UNDEX) has been experimentally and computationally investigated. The problem analyzed consists of a flat steel plate adjacent to an explosive within an underwater environment. The experiments were performed for different standoff distances using high speed photography to capture the behavior of the underwater explosive gas bubble, and a series of pressure transducers, to record the emitted pressure histories. The numerical simulations were performed with the fully coupled Eulerian–Lagrangian fluid structure interaction code DYSMAS. The numerical simulations were validated with the experiments in terms of the detonation pressure, structural surface pressures, and UNDEX gas bubble growth and collapse. Results show that the UNDEX standoff distance greatly influences the gas bubble's shape, migration speed, jetting behavior, and loading into its adjacent structure.

Keywords

UNDEX , Gas bubble , Shockwave , Eulerian–Lagrangian

3.1 Introduction

Explosive characteristics in the free-field (no surrounding structures) underwater environment such as the pressure, decay, and bubble behavior are well understood. During detonation, the explosive material will produce extreme pressures and gaseous byproducts that generates a gas bubble [1]. The behavior of the bubble depends on the depth of detonation and on nearby structures [2]. This research arises from the concern of damage to marine structures from underwater explosive (UNDEX) loading, such as ships, oil pipes, and underwater vehicles. In undesirable circumstances, these structures can be subjected to shock and blast loadings which will lead to gas bubbles in an underwater environment. To avoid damages and catastrophic losses under these highly dynamic conditions the bubble to structure interaction needs to be properly characterized experimentally and predicted through computational means.

C. Javier (✉) · M. Galuska · M. Papa · J. LeBlanc
Naval Undersea Warfare Center (Division Newport), Newport, RI, USA
e-mail: carlos.r.javier@navy.mil

H. Matos · A. Shukla
Naval Undersea Warfare Center (Division Newport), Newport, RI, USA

Dynamic Photo Mechanics Laboratory, Department of Mechanical, Industrial and Systems Engineering, University of Rhode Island, Kingston, RI, USA

J. Notbohm et al. (eds.), *Mechanics of Biological Systems and Materials & Micro-and Nanomechanics & Research Applications*, Conference Proceedings of the Society for Experimental Mechanics Series,
https://doi.org/10.1007/978-3-030-59765-8_3

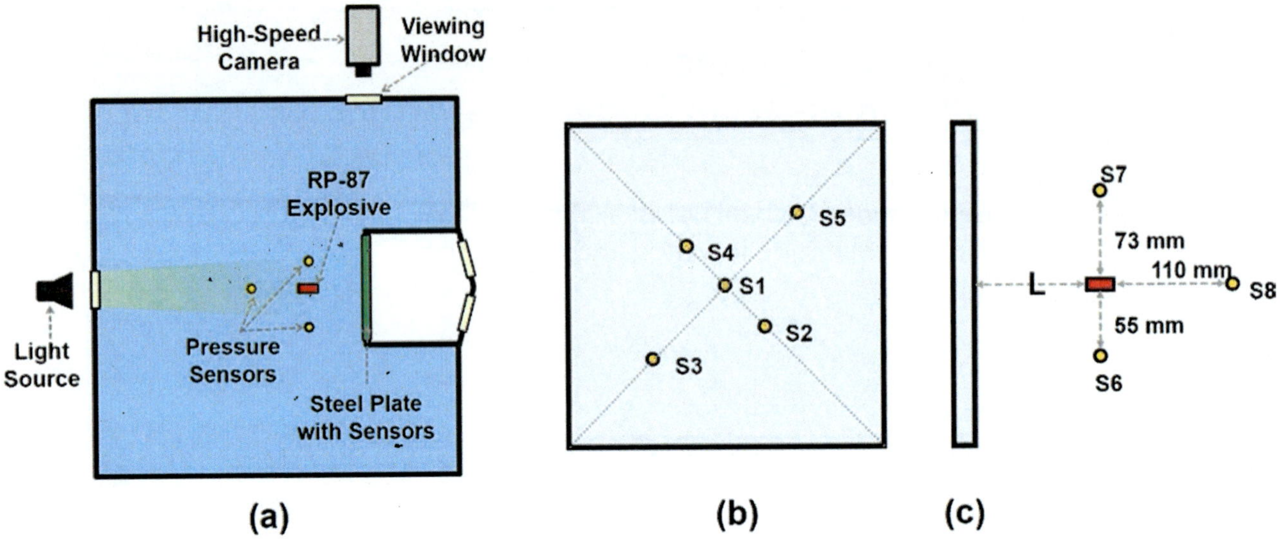

Fig. 3.1 (**a**) Top view of the UNDEX experimental facility with the explosive located 530 mm below the water line (**b**) Schematic of the steel plate with the location of the mounted pressure sensors (**c**) Distance between the explosive and the water pressure sensors

3.2 Experimental Procedure and Finite Element Approach

The experiments were performed using a 13 mm thick steel shell UNDEX facility with dimensions of 1.2 m × 1.2 m × 1.2 m.

A steel plate was fully clamped to an air-backed chamber; the steel plate was low carbon steel with length and width of 305 × 305 mm and a thickness of 13 mm. An RP-87 explosive was placed at the midspan of the steel plate as shown in Fig. 3.1a, b. A Photron SA1 by Photron USA cameras was placed normal to a viewing port located on the tank for visualizing the UNDEX gas bubble dynamics at 10,000 frames per second.

Three PCB 138A05 dynamic pressure transducers were used to record the emitted signatures at 2 MHz in water. To record the loading on the steel plate, five PCB102A pressure transducers were mounted flush to the plate, as shown in Fig. 3.1b. Three standoff distances between the explosive and the steel plate were selected based on the maximum radius achieved by the gas bubble. These standoff distances are: (1) less than the maximum bubble radius (L = 55 mm); (2) equivalent to the maximum bubble radius (L = 73 mm); and (3) greater than the bubble radius (L = 110 mm). The dimensionless standoff parameter γ was utilized such that

$$\gamma = \frac{L}{R_{\max}} \tag{3.1}$$

The value for γ obtained from Eq. (3.1) for the three standoff distances were: $\gamma = 1.50$, $\gamma = 1.00$ and $\gamma = 0.75$. Computational modeling of the UNDEX experiments was made with the Dynamic System Mechanics Advanced Simulation (DYSMAS) software. The computer simulations modeled the detonation of the RP-87 charge including the resulting pressures and wave propagation, the fluid structure interaction phenomena between the pressure front and the steel plate, and expansion and subsequent collapse of the gas bubble.

3.3 Results

The pressure generated by the initial shockwave for $\gamma = 1.50$ is shown in Fig. 3.2a. The naming scheme of the sensors shown are taken from Fig. 3.1b.

Considering S5 for $\gamma = 1.50$, the distance from the explosive to the sensor was 135 mm. In terms of sensor S5, the pressure decays with *1/R* as in the case in a free field environment. For the case of $\gamma = 1.00$, the shockwave pressures for S1and S5 are shown in Fig. 3.2b. The measured distance between the explosive and sensor S5 was 100 mm. It should be noted that the pressure does not decay with *1/R* for $\gamma = 1.00$ and $\gamma = 0.75$ (Fig. 3.2c).

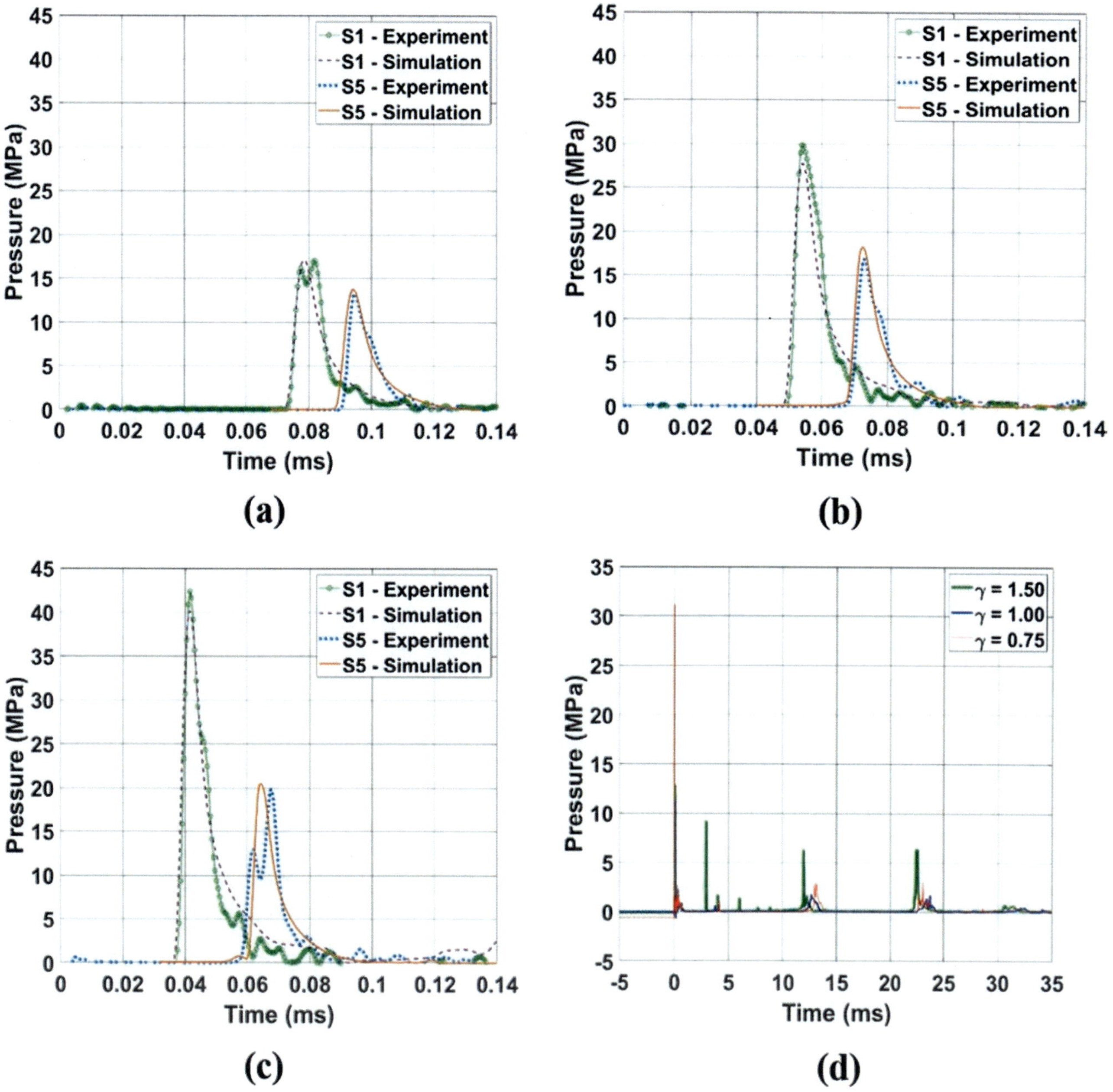

Fig. 3.2 (**a**) Experimental and simulated shock wave about the surface of the steel plate for sensors S1 and S5 for $\gamma = 1.50$ (**b**) Experimental and simulated shock wave about the surface of the steel plate for sensors S1, and S5 for $\gamma = 1.00$ (**c**) Experimental and simulated shock wave about the surface of the steel plate for sensors S1 and S5 for $\gamma = 0.75$ (**d**) Experimental pressure-time history for $\gamma = 1.50$, $\gamma = 1.00$, and $\gamma = 0.75$ recorded by sensor S1

When the shockwave initially impinges on the structure (S1), the shockwave is reflected back onto the remainder of spherically shaped shockwave, which alters the magnitude of the incoming wave as it impinges on the remainder of the plate, therefore utilizing the pressure decay of *1/R* is not sufficient in determining the magnitude throughout the rest of the plate. The pressure data for all standoff distances is shown in Fig. 3.2d. The first and second bubble collapses occur at about ~12 ms, and 23 ms, respectively for all standoff distances. To evaluate the damage potential of the shock wave and the subsequent bubble pulses, both the peak pressures and the duration are of equal importance, therefore calculating the total impulse generated by each event is necessary. The impulses for $\gamma = 1.50$, $\gamma = 1.00$ and $\gamma = 0.75$ are tabulated in Table 3.1 both for the experimental and simulation results. It should be noted that the second bubble pulse is not included for the simulation results. For the case of $\gamma = 1.50$, the experimental shockwave impulse was 190 Pa*s, with the first bubble pulse being about 6 times higher at 1050 Pa*s. The impulse of the second bubble pulse was 1650 Pa*s, about 1.6 times higher than the first

Table 3.1 Tabulated impulse results, both experimental and simulated values for the cases of $\gamma = 1.50$, 1.00, and 0.75. The experimental values are an average of the two experiments conducted for each case

Case	Event	Experiment impulse (Pa*s)	Simulation impulse (Pa*s)
$\gamma = 1.50$	Shockwave	190	205
	1st Bubble	1030	1170
	2nd Bubble	1650	–
$\gamma = 1.00$	Shockwave	360	345
	1st Bubble	3430	2530
	2nd Bubble	990	–
$\gamma = 0.75$	Shockwave	415	465
	1st Bubble	3070	2350
	2nd Bubble	790	–

bubble pulse, since the first bubble occurs away from the plate and the second bubble pulse occurs on the plate. The maximum percent difference between the experimental and numerical shockwave impulses were 7.9%, indicating good correlation. The maximum percent error between experiment and simulation increases in terms of the impulse of the first bubble is about 14%. The experimental shockwave impulse for $\gamma = 1.00$ was taken to be 360 Pa*s, with a 4.3% error between experiment and simulation. Similar to the $\gamma = 1.50$ case, the shock wave impulse was small relative to the impulse of the first bubble pulse, which is almost 10 times larger than the shock wave impulse (3430 Pa*s). Moreover, the impulse obtained by the simulation for the first bubble was 2530 Pa*s, which gives a maximum difference error of 35.6%. In this case, the simulation did not predict the magnitude of the bubble, or the duration, therefore a large error in impulse between the experiment and simulation is shown. The impulse of the second bubble pulse was 1050 Pa*s, about 2.5 times lower than the first bubble pulse. In this case, the first bubble had sufficient distance from the rigid structure and was able to achieve 70% of the volume as if it were in a free field environment, with a semi-spherical shape, therefore the first bubble was able to generate an impulse much larger than the shockwave and the second bubble pulse (in $\gamma = 1.50$, the second bubble generated a higher impulse than the first). As the bubble collapses and dissipates energy, the impulse decreases, highlighted as a 65% drop from the second bubble pulse to the third. For $\gamma = 0.75$, the shock wave impulse was 406 Pa*s.

The impulse of the first bubble was 3070 Pa*s, lower than the $\gamma = 1.00$ case. In the $\gamma = 0.75$ case, the bubble growth was impeded by the rigid structure, reaching 68% of the maximum bubble volume for a free field environment. Moreover, the shape of the bubble was distorted and did not have a symmetric collapse, potentially decreasing the overall impulse generated. The impulse of the second bubble pulse was 834 Pa*s, again lower than the second bubble pulse of the $\gamma = 1.00$ case. The bubble pulse of the third bubble experiments and simulation was 615 Pa*s.

Acknowledgements The authors would kindly like to acknowledge the National Institute for Undersea Vehicle Technology (NIUVT) for leading the collaboration effort between the University of Rhode Island, and the Naval Undersea Warfare Center Division Newport.

References

1. Cole, R.H.: Underwater explosions. Princeton University Press, Princeton, NJ (1948)
2. Chapman, R.B.: Measurement of the waveform parameters of shallow explosive charges. J. Acoust. Soc. Am. **78**, 672 (1985). https://doi.org/10.1121/1.392436

Optimization for Improved Energy Absorption and the Effect of Density Gradation in Cellular Materials

4

Vijendra Gupta, Dennis Miller, Addis Kidane, and Michael Sutton

Abstract

Cellular foam materials are used increasingly in a wide range of applications involving energy absorption and impact protection as they are capable of absorbing a large amount of energy without a significant increase in load. In this study, the optimum relative density of cellular material that maximizes the energy absorption performance is investigated for low impact velocity applications. Idealized material models are employed to represent the material behavior at low impact velocities. Quasi-static compression experiments covering a wide range of densities are performed to extract material model parameters for rigid polyurethane foam. Using the material models with experimentally determined parameters, the optimal density is identified and the relative increase in energy absorption of the functionally graded material over a uniform density material is determined. It is shown that the uniform density configuration has a superior energy performance. It is also shown that density gradation is advantageous up to a certain value of average relative density, designated as the critical average relative density. This critical average relative density is shown to be 0.2 for rigid polyurethane foam.

Keywords

Functionally graded foam , Energy absorption , Cellular materials , Rigid polyurethane , Density gradation

4.1 Introduction

Cellular foam materials possess exceptional energy absorption capacity. In rigid polymer foams, the primary mechanism of energy absorption is plastic yielding. Yielding occurs at low-stress values, with large deformation leading to substantial energy absorption. The relative density of the cellular material largely governs mechanical response and is the primary variable for density gradation. Energy absorption diagrams have been used to study energy absorption as a function of relative density [1, 2]. In the present study, idealized constitutive models are used in order to model the mechanical behavior of the functionally graded foam materials. The functional gradation of properties of cellular materials provides opportunities to tailor the material as per the design requirements. The effect of nonlinear density gradation on the energy absorption of foam is investigated theoretically in [3]. The authors in [4] explored the energy absorption performance of discretely graded foam experimentally and used a semi-analytical approach for modeling continuously graded foams. In [5], the investigators proposed a nonlinear constitutive phenomenological model to capture the complex behavior of foam under large compressive loads. There have also been several numerical studies based on finite element methods [6–9]. Some studies used continuum-based models [6, 7] while others used cell-based models [8, 9]. The investigators, in [6] modeled various gradient functions of density to improve energy absorption performance. Functionally graded foams with linear gradation were used in [8] to

V. Gupta (✉) · D. Miller · A. Kidane · M. Sutton
Department of Mechanical Engineering, University of South Carolina, Columbia, SC, USA
e-mail: vijendra@email.sc.edu

J. Notbohm et al. (eds.), *Mechanics of Biological Systems and Materials & Micro-and Nanomechanics & Research Applications*, Conference Proceedings of the Society for Experimental Mechanics Series,
https://doi.org/10.1007/978-3-030-59765-8_4

investigate the effect of irregularities due to a missing cell cluster. Functional gradation was studied in [9] using polynomial variations in the number of pores.

In this study, the energy absorption performance of a functionally graded cellular material subjected to low-velocity deformation applications is investigated using idealized constitutive models. Quasi-static experiments are performed on rigid polyurethane (PUR) foam to obtain the material parameters required for the model. The functional gradation is modeled using power-law variations of relative density. Energy absorbed per unit volume and energy absorbed per unit mass are used as performance metrics. The comparative advantage of the functionally graded material over uniform density foam material is also studied.

4.2 Uniform Density Model

Cellular materials have characteristic mechanical behavior with three distinct regimes: linear elastic; plateau; and densification. The stress is proportional to strain in the linear elastic region. The plateau region is the key region where significant energy absorption takes place at approximately constant stress. The densification stage is marked by a sharp increase in stress as the material densifies. The typical mechanical behavior of cellular foam is shown in Fig. 4.1. The foam behavior can be idealized as having constant stress followed by a rigid behavior as shown in the figure. The strain at which the material becomes rigid is referred to as the densification strain. The idealized behavior captures the main feature, which is the plateau region, of the foam response while compromising on other less important elements. This model has also been used to study the dynamic response of wood and is known as rigid, perfectly-plastic, locking (RPPL) model [3]. Only two parameters are needed to completely define this idealized model viz., plateau stress (σ_p) and densification strain (ϵ_d). These parameters, and in general the stress-strain response, mainly depend on the relative density ($\bar{\rho}$) which is defined as the ratio of density of cellular material (ρ) and the density of the parent material (ρ_S). The plateau stress and densification strain can be easily identified for various relative densities of the foam from the empirically obtained stress-strain curves.

It is known from the micro-mechanical model that the plateau stress is proportional to the square of relative density. For more refined models, there are additional terms to describe the behavior [1]. However, in general, the plateau stress can be described as having power-law variation with relative density:

$$\sigma_p = C_1 \sigma_S \bar{\rho}^n \tag{4.1}$$

where σ_S represents yield stress of the solid material, n is the power-law exponent and C_1 is a non-dimensional material constant. Densification strain is modeled using a negative linear relationship with relative density and is given by,

$$\epsilon_d = 1 - C_2 \bar{\rho} \tag{4.2}$$

where C_2 is a non-dimensional parent material-dependent constant. The energy absorbed per unit volume (E_v) is given by the area under the stress-strain curve. For constant stress, it is given by,

$$E_v = \sigma_p \epsilon_d \tag{4.3}$$

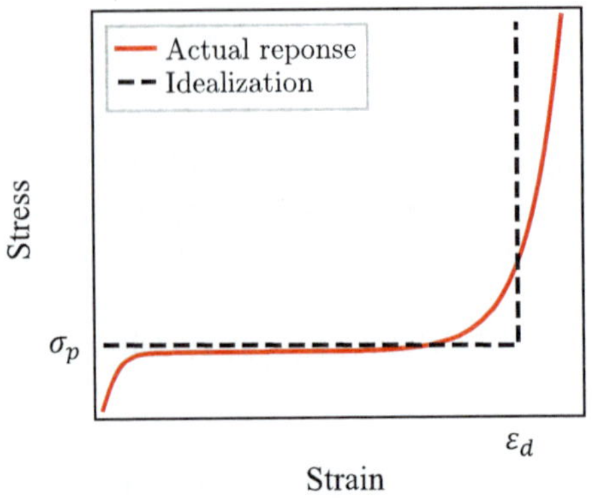

Fig. 4.1 Mechanical response of foam

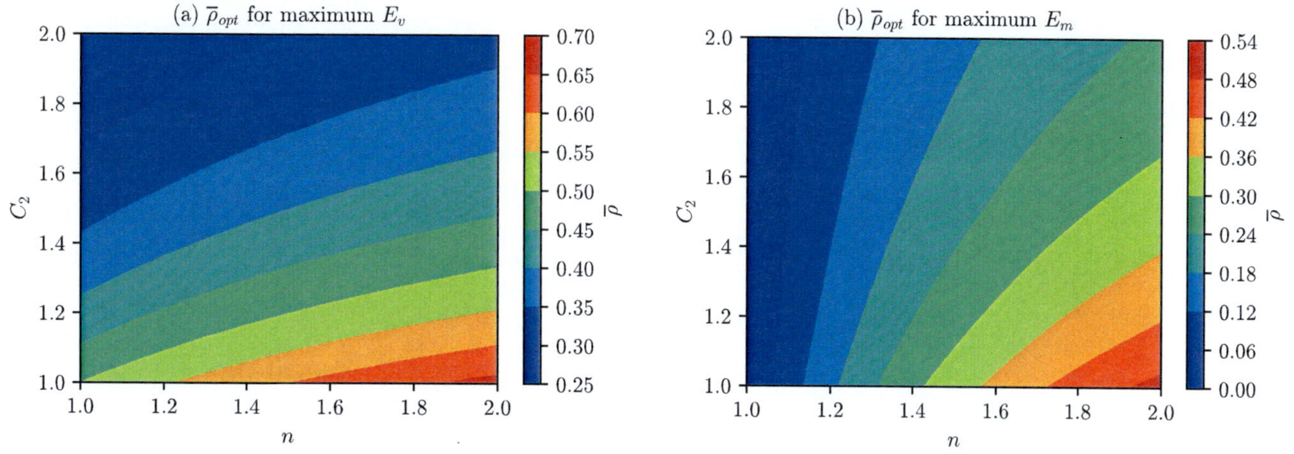

Fig. 4.2 Optimum uniform density for (**a**) maximum energy per unit volume (**b**) maximum energy per unit mass

Energy absorbed per unit mass can be used as a metric where the mass also needs to be minimized along with the maximum energy absorption. The energy absorbed per unit mass can be obtained by dividing E_v by mass of the foam and is given by,

$$E_m = \frac{\sigma_p \epsilon_d}{\bar{\rho}\rho_s} \tag{4.4}$$

The optimum values of relative density can be obtained by differentiating Eqs. (4.3) and (4.4), and using Eqs. (4.1) and (4.2). The optimum values are given by,

$$\bar{\rho}_{opt(v)} = \frac{n}{C_2(n+1)} \tag{4.5}$$

$$\bar{\rho}_{opt(m)} = \frac{n-1}{C_2 n} \tag{4.6}$$

where $\bar{\rho}_{opt(v)}$ and $\bar{\rho}_{opt(m)}$ represent the optimal relative density for maximum energy absorbed per unit volume and maximum energy absorbed per unit mass, respectively. It is seen from Eqs. (4.5) and (4.6) that the optimum density depends upon the material constants C_2 and n but is independent of C_1. The contour plots for optimal relative density are shown in Fig. 4.2. The optimum relative density increases with increase in n but decreases with increase in C_2 for both maximum E_v and maximum E_m. Once the material constants C_2 and n are known, the appropriate optimal relative density of foam can be selected from the plots.

4.3 Uniform Density Experiments

The material constants C_2 and n that govern the optimal energy absorption performance are determined for PUR foam through quasi-static compression experiments. LAST-A-FOAM® FR-7100 Series PUR obtained from General Plastics Manufacturing Company is considered for this study. The properties of the parent solid material PUR were obtained from [1]. The density, elastic modulus and yield strength of the solid material PUR are 1200 kg/m^3, 1600 MPa and 127 MPa, respectively. The relative densities of the specimens were experimentally computed from the volume and weight measurements. The samples were machined to a nominal size of 15 × 15 × 15 mm^3 from the foam blocks. The details are given in Table 4.1. The mean values of the size dimensions and weights are reported. It is found that the actual relative densities of some of the variants differ from the nominal values specified by the manufacturer.

Uniform density foam samples of different relative densities are tested at room temperature using an electromechanical testing machine (MTS Exceed Model E43.504). As the sample is compressed, it is confined on all sides in a relatively rigid cell made up of aluminum and 12.5 mm thick polycarbonate plate on the front side (shown in Fig. 4.3). The applied compres-

Table 4.1 Density measurements

Nominal density (kg/m^3)	Length (mm)	Width (mm)	Thickness (mm)	Weight (g)	Actual density (kg/m^3)	Actual relative density (%)
64	14.78	15.00	14.93	0.21	64.9	5.4
96	14.83	14.95	14.96	0.31	93.3	7.8
128	14.56	15.01	15.05	0.41	124.7	10.4
160	15.01	14.91	15.00	0.51	152.6	12.7
320	15.00	15.02	14.99	1.06	314.8	26.2
480	15.01	15.00	15.00	1.57	464.4	38.7
640	15.03	15.00	14.98	2.34	693.7	57.8

Fig. 4.3 Cellular material in a confined cell

sive load was obtained from the built-in 30 kN force transducer and the strains were obtained from the crosshead displacement. The experiments were performed at a strain rate of $1.67 \times 10^{-3}\ s^{-1}$ which is a standard strain rate used for quasi-static compression tests as mentioned in [10]. The stress-strain response for different relative densities obtained from the tests is shown in Fig. 4.4.

The variation of plateau stress and densification strain with relative density is shown in Fig. 4.5. The densification strain shown here represents the strain at the onset of the densification. Different researchers have used different definitions for the densification strain of cellular materials. In [1], the authors defined the densification strain based on the initial relative density of the material that lacks a distinct plateau region. This requires the knowledge of the lowest relative density which does not possess the plateau region for obtaining densification strain. In [11], the investigators used the intersection of tangents to the stress-strain curve to define densification strain in wood, which is also a class of cellular materials. This method is subjective since the point of tangency is not clearly defined. The slope of the stress-strain curve is continuously changing, and different points used for tangents will lead to different densification strains. The authors in [12] suggested the use of the strain corresponding to maximum efficiency for uniquely defining the onset of densification. The efficiency is defined as the ratio of energy absorbed per unit volume and applied stress. This method is useful for low relative densities since they have a constant slope in the plateau region and are characterized by a sharp increase in stress at the beginning of the densification phase. However, for higher relative density material (see Fig. 4.4b), the plateau region has a positive slope and the stress increases more gradually. In this case, the efficiency method will tend to give higher densification strains. In the current study, the onset of the densification strain is defined as the strain at which the tangent modulus reaches 10% of the elastic modulus of the solid material. The material will be fully densified when the tangent modulus approaches the elastic modulus of the solid material. Also, it has a finite initial density before any load applied. So, a small percentage of elastic modulus of the solid material should be a good criterion for densification. Considering that the higher relative density material has a positive slope in the plateau region, a value closer to but greater than this slope is chosen as the criterion for onset of densification. Once

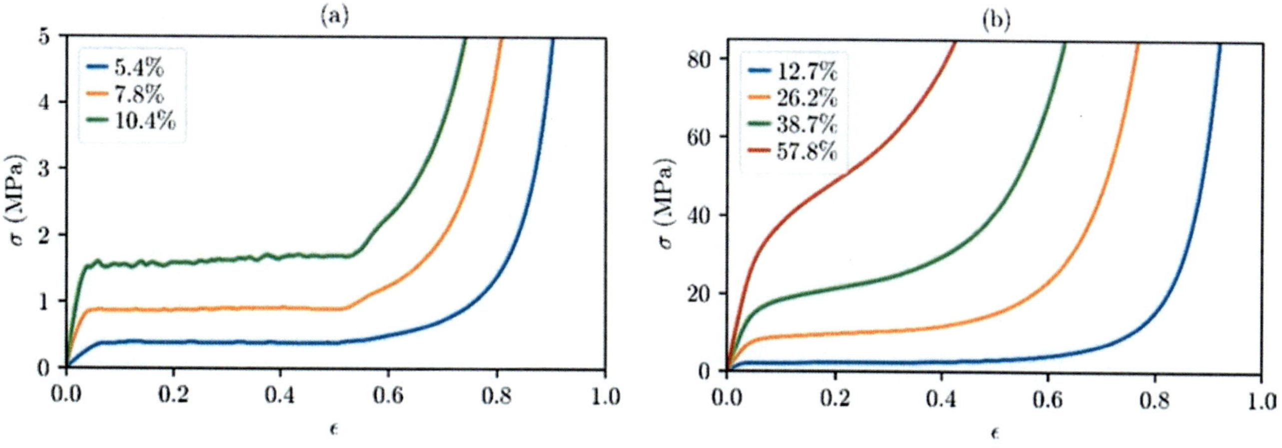

Fig. 4.4 Stress-strain response (**a**) low relative densities (**b**) high relative densities

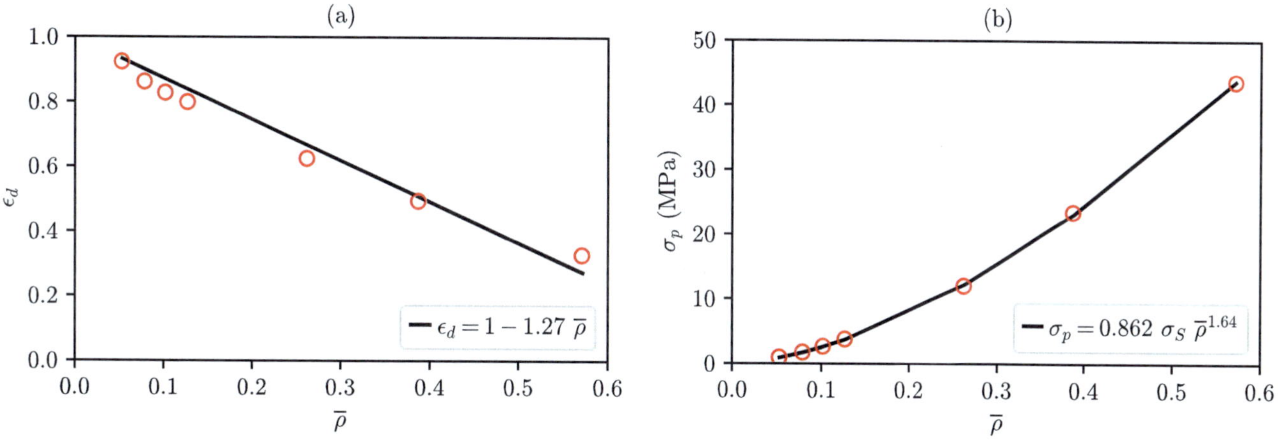

Fig. 4.5 (**a**) Densification strain variation with density (**b**) Plateau stress variation with density

the densification strain is determined, the plateau stress can be defined as the average value of the stress up to the densification strain and is given by the following equation:

$$\sigma_p = \frac{\int_0^{\epsilon_d} \sigma d\epsilon}{\epsilon_d} \tag{4.9}$$

The material constants are obtained by fitting the Eqs. (4.1) and (4.2) to the stress-strain data. The material constants C_1, C_2 and n are found to be 0.862, 1.27 and 1.64, respectively. Thus, using Eqs. (4.5) and (4.6), the optimal relative density for maximum energy per unit volume and maximum energy absorbed per unit mass are determined to be 48.9% and 30.7%.

4.4 Functionally Graded Model

The relative density is assumed to vary continuously in a functionally graded material considered in this study. The relative density ($\bar{\rho}$) at any location (X) along the length of the foam (L) is modeled using the following power law equation,

$$\bar{\rho} = \bar{\rho}_1 + (\bar{\rho}_2 - \bar{\rho}_1)\left(\frac{X}{L}\right)^m \tag{4.10}$$

where $\bar{\rho}_1$ and $\bar{\rho}_2$ are relative densities at the two ends of the graded foam. m represents the exponent of the power-law variation. It is to be noted that since relative density is varying along the length, the plateau stress and densification strain will also vary accordingly. The average relative density ($\bar{\rho}_0$) of the specimen can be computed using the following equation,

$$\bar{\rho}_0 = \bar{\rho}_1 + \frac{\bar{\rho}_2 - \bar{\rho}_1}{m+1} \tag{4.11}$$

The relative density at any point can be expressed in terms of average density $\bar{\rho}_0$ and density difference $\Delta\bar{\rho}$ between the two ends of the cellular material. By rearranging Eq. (4.10), the relative density can be written as

$$\bar{\rho} = \bar{\rho}_0 \left[1 - \frac{\Delta\bar{\rho}}{\bar{\rho}_0 (m+1)} + \frac{\Delta\bar{\rho}}{\bar{\rho}_0} \left(\frac{X}{L} \right)^m \right] \tag{4.12}$$

Positive $\Delta\bar{\rho}$ indicates that the density is increasing and $\bar{\rho}_2 > \bar{\rho}_1$ while negative $\Delta\bar{\rho}$ indicates that the density is decreasing and $\bar{\rho}_1 > \bar{\rho}_2$. Energy absorbed by the material is given by the area under the stress-strain curve. Since the density is varying along the length, the energy absorbed per unit volume (E_v) and energy absorbed per unit mass (E_m) for functionally graded material can be defined by the following equations,

$$E_v = \frac{1}{L} \int_0^L \sigma_p (X) \epsilon_d (X) dX \tag{4.13}$$

$$E_m = \frac{1}{\bar{\rho}_0 \rho_s L} \int_0^L \sigma_p (X) \epsilon_d (X) dX \tag{4.14}$$

In order to determine the relative advantage of the functionally graded material over uniform density material, the energy ratio ($\bar{E}$) can be defined as the ratio of the energy absorbed by the graded material over energy absorbed by the uniform density material.

$$\bar{E} = \frac{E_v (graded)}{E_v (uniform)} \tag{4.15}$$

E_v and E_m are computed numerically for PUR using the material constants $C_1 = 0.862$, $C_2 = 1.27$ and $n = 1.64$ obtained in the previous section. Contour plots of E_v and E_m, showing their dependence on average relative density and relative density gradient, are plotted in Fig. 4.6. The ratio, $\Delta\bar{\rho} / \bar{\rho}_0$ represents the normalized density gradient and the line corresponding to $\Delta\bar{\rho} / \bar{\rho}_0 = 0$ in the figure represents the case of uniform density. The density gradient increases away from this line. The response is independent of the direction of the density gradient because force equilibrium always holds at quasi-static loading rates. Hence the line is ought to be symmetric about $\Delta\bar{\rho} / \bar{\rho}_0 = 0$. It is seen that E_v and E_m are maximum at different average relative density but both the optimal points lie on the $\Delta\bar{\rho} / \bar{\rho}_0 = 0$. In this study, E_v is maximum at $\bar{\rho}_0$ = 48.9%, while E_m is maximum at $\bar{\rho}_0$ = 30.7% (highlighted with black dot in Fig. 4.6). These are the same optimal values obtained in the previous section. Thus, this indicates that the uniform density distribution is the optimal configuration for better energy absorption performance. Functionally grading the material or changing the relative density decreases the energy absorption performance. The energy ratio $\bar{E}$ compares the performance of functionally graded material to the uniform density material. The contour plot of the energy ratio is shown in Fig. 4.7. It is seen that the energy ratio is higher at lower average relative densities and higher density gradients. On the contrary, the energy ratio decreases when both the average relative density and density gradient is high. It is interesting to note that the energy ratio is quasi-constant at a critical value of $\bar{\rho}_0$ = 0.2. Below this value of average relative density, the energy ratio of functionally graded material is better than for a uniform density material. On the other hand, above the critical value, functional gradation reduces the energy ratio.

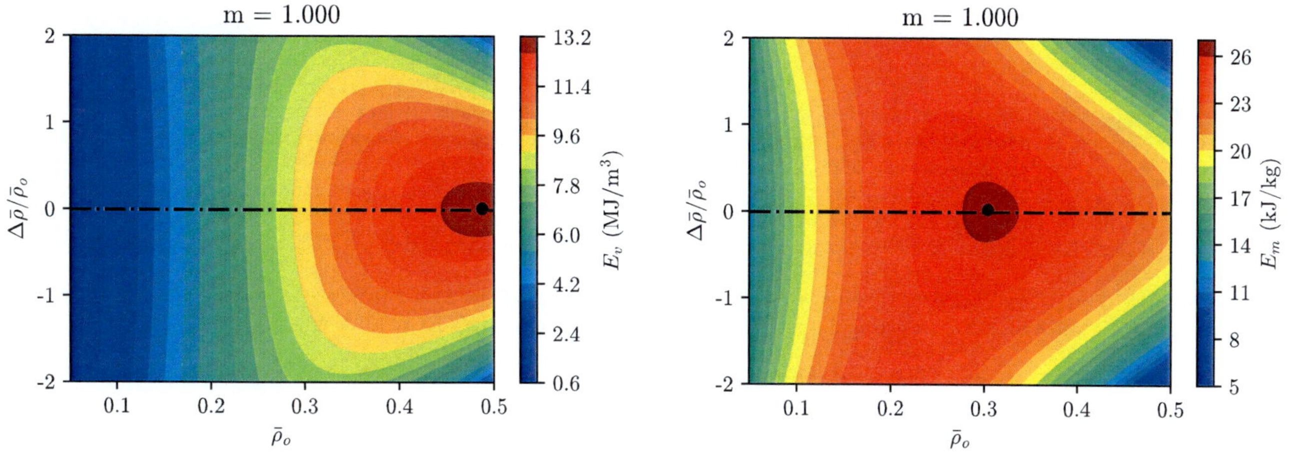

Fig. 4.6 (**a**) Energy absorbed per unit volume and (**b**) Energy absorbed per unit mass of functionally graded cellular material

Fig. 4.7 Energy ratio as a function of average relative density and density gradient

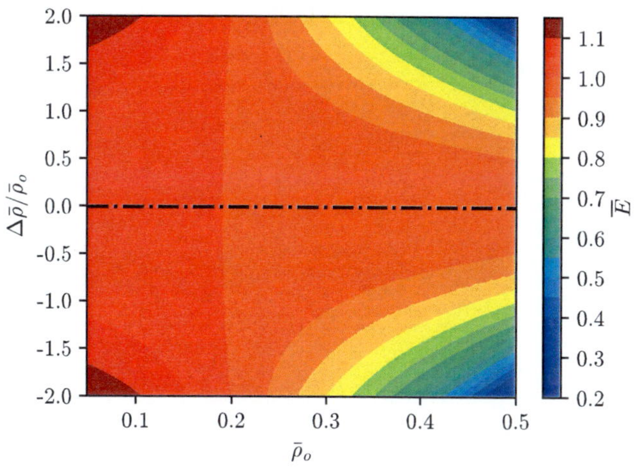

4.5 Conclusions

The current study explores the optimal density configuration of PUR foam using idealized constitutive models. The optimal density is determined through analytical modeling for the maximum energy absorption performance of the cellular material. Quasi-static compressive experiments are performed to determine the material constants for PUR. Energy absorbed per unit volume and the energy absorbed per unit mass are shown to be maximum at the relative density of 48.9% and 30.7%, respectively. The critical value of average relative density that determines the relative advantage of functional gradation is shown to be 0.2 for PUR. Although the uniform density distribution is demonstrated to be optimal in quasi-static applications, the density variation in the material allows designers to have more flexibility and enhance other aspects of the performance for the material. Thus, it is important to understand the effect of functional gradation on energy absorption performance. Results from this study will be useful in the design of functionally graded material for relatively low loading rates. Future studies will include nonlinear models to capture the response more accurately. The study also can be extended to incorporate the dynamic response of cellular materials when subjected to higher impact velocities.

Acknowledgments Financial support provided by the US Army Research Office via grant W911NF-17-S-0002 is gratefully acknowledged.

References

1. Gibson, L.J., Ashby, M.F.: Cellular solids: Structure and properties, 2nd edn. Cambridge University Press, Cambridge (1997)
2. Avalle, M., Belingardi, G., Montanini, R.: Characterization of polymeric structural foams under compressive impact loading by means of energy-absorption diagram. Int. J. Impact Eng. **25**(5), 455–472 (2001)
3. Liu, H., Zhang, Z., Liu, H., Yang, J., Lin, H.: Theoretical investigation on impact resistance and energy absorption of foams with nonlinearly varying density. Compos. Part B Eng. **116**, 76–88 (2017)
4. Koohbor, B., Kidane, A.: Design optimization of continuously and discretely graded foam materials for efficient energy absorption. Mater. Des. **102**, 151–161 (2016)
5. Liu, Q., Subhash, G.: A phenomenological constitutive model for foams under large deformations. Polym. Eng. Sci. **44**(3), 463–473 (2004)
6. Cui, L., Kiernan, S., Gilchrist, M.D.: Designing the energy absorption capacity of functionally graded foam materials. Mater. Sci. Eng. A. **507**(1–2), 215–225 (2009)
7. Kiernan, S., Gilchrist, M.D.: Towards a virtual functionally graded foam: Defining the large strain constitutive response of an isotropic closed cell polymeric cellular solid. Int. J. Eng. Sci. **48**(11), 1373–1386 (Nov. 2010)
8. Ajdari, A., Nayeb-Hashemi, H., Vaziri, A.: Dynamic crushing and energy absorption of regular, irregular and functionally graded cellular structures. Int. J. Solids Struct. **48**(3–4), 506–516 (2011)
9. Chen, D., Kitipornchai, S., Yang, J.: Dynamic response and energy absorption of functionally graded porous structures. Mater. Des. **140**, 473–487 (2018)
10. ASTM D1621-16: Standard test method for compressive properties of rigid cellular plastics. ASTM International, West Conshohocken, PA (2016)
11. Reid, S., Peng, C.: Dynamic uniaxial crushing of wood. Int. J. Impact Eng. **19**(5–6), 531–570 (1997)
12. Li, Q.M., Magkiriadis, I., Harrigan, J.J.: Compressive strain at the onset of densification of cellular solids. J. Cell. Plast. **42**(5), 371–392 (2006)

5

Quantifying Ultrasonic Deformation of Cell Membranes with Ultra-High-Speed Imaging

A. Marek, A. De Grazia, X. Régal, D. Carugo, and F. Pierron

Abstract

We present a new method for controllable loading of cell models in an ultrasonic (20 kHz) regime. The protocol is based on the inertial-based ultrasonic shaking test and allows to deform cells in the range of few mm/m to help understand potential consequences of repeated loading characteristic of ultrasonic cutting.

Keywords

Ultra-high speed imaging , Ultrasound , GUVs , IBUS , Mechanobiology

5.1 Introduction

Ultrasonically-powered cutting tools are a promising alternative for cutting hard tissues with higher precision and lower damage compared to the currently employed tools. During a surgical procedure a fresh scalpel causes cell deaths in a nearby region called a 'zone of cell death' (several hundreds of micrometers wide) slowing down the healing process [1]. Ultrasonic cutting tools have a potential to reduce/control the width of the zone of cell death. However, the interaction between the ultrasonic loading and cell death is still not well established.

The oscillating cutting blade interacts with the cutting sites and produces repeated deformation that travels through the extracellular matrix (ECM) in the form of mechanical waves. As the ECM deforms, so do the cells attached to it which in turns can lead to activation of molecular pathways leading to death, or potentially promote healing as used in low intensity pulsed ultrasound (LIPUS) treatments [2–4]. Due to high frequency nature of the loading it is very difficult to reproduce *in vitro* the dynamic loading that is relatable to what the cells experience *in vivo*.

The most common experimental techniques to study biological effects of ultrasonic loading on cellular entities are cavitation tests (often leading to sonoporation) [5, 6] and direct exposure to acoustic waves in liquid media [7, 8]. The former represents much more violent stimulus, not necessarily applicable to the cutting scenario, while the latter makes oversimplified assumptions regarding acoustic conditions (pressure) such as lack of reflections, no formation of standing waves etc. and does not measure the deformation of cells directly.

In this contribution we propose a test based on the inertial-based ultrasonic shaking test (IBUS) [9] in which cells are attached to a substrate deforming with ultrasonic frequency, thus transferring the loading onto them. For this task we use a

A. Marek (✉) · A. De Grazia · X. Régal · F. Pierron
Faculty of Engineering and Physical Sciences, University of Southampton, Southampton, UK
e-mail: a.marek@soton.ac.uk

D. Carugo
Department of Pharmaceutics, UCL School of Pharmacy, University College London (UCL), London, UK

J. Notbohm et al. (eds.), *Mechanics of Biological Systems and Materials & Micro-and Nanomechanics & Research Applications*, Conference Proceedings of the Society for Experimental Mechanics Series,
https://doi.org/10.1007/978-3-030-59765-8_5

cell membrane model, giant unilamellar vesicles (GUVs) that consist of lipid bilayer enclosing an aqueous solution. The substrate is made out of clear Poly(methyl methacrylate) (PMMA) and allows to image the deforming cells in real time using an ultra-high speed camera (Shimadzu HPV-X).

We demonstrate that the attached GUVs deform with the same frequency and strain amplitude as the substrate calibrated using the IBUS test and so the test offers an interesting platform to study biological effects of attached cells under known amplitude of ultrasonic deformation. In future, this platform could be used to study fatigue of cell membranes, as well as biological responses of living cells to a known history of mechanical/thermal stimuli.

5.2 Methods

The protocol consists of two steps: preparing and calibrating the substrate according to the IBUS test and then imaging deformation of attached GUVs under a bright field microscope setup.

In the first step, clear PMMA coupons (55 × 20 × 3 mm) were laser-cut from a 3 mm thick sheet (Perspex) and tested following the IBUS procedure. The coupon's length was selected such that its first longitudinal mode corresponds to 20 kHz excitation. One face of the coupon was covered with white acrylic paint and printed with a regular pattern of grid (0.82 mm pitch) to measure full-field displacements using the grid method (Fig. 5.1) [10], while the other side was covered with black acrylic paint to facilitate temperature measurements. The coupon was glued to a sonotrode (SinapTec NexTgen) that excites one of the edges at 20 kHz. The sonotrode was programmed to oscillate at a known fraction of its maximal amplitude and the corresponding deformation was measured using an ultra-high speed camera (Shimadzu HPV-X) recording at 500 kfps. From these measurements it was possible to calibrate amplitude of longitudinal strain at each of the sections of the specimen with the sonotrode power, avoiding the need for direct measurements in subsequent tests.

The second step consists of preparing GUVs and functionalizing the PMMA coupons such that they can be used to transfer the ultrasonic loading onto the vesicles. First, an O-ring was attached at the middle of the coupon using a silicone gel so that the stiffness (and thus the resonance response) of the coupon is not altered (Fig. 5.2a). The O-ring serves as a well to contain liquid solutions needed to keep and attach GUVs on the substrate. To bio-functionalize the PMMA surface and allow for a good attachment of GUVs to it an avidin-biotin link was utilized. The coupon was cleaned with ethanol and the well was washed with phosphate buffer solution (PBS) twice. Then, the well was filled with approximately 250 μl of biotinylated bovine serum albumin (Pierce™, Thermo Scientific) and incubated at room temperature for 1 h. Then, the well was washed twice with PBS and filled with 250 μl of avidin solution and incubated for another 2 h. Afterwards, the well was washed twice with PBS and the coupon was glued to the sonotrode using a cyanoacrylate glue (Loctite 422) and was ready to be filled with GUVs suspension.

The GUVs were then prepared according to the electroformation protocol presented by Pereno et al. [11]. Two steel needles were covered with a lipid solution of DOPC and 1,2-dipalmitoyl-sn-glycero-3-phosphoethanolamine-N-(cap biotinyl) (sodium salt) (molar ratio of 10:1) dissolved in chloroform to the final concentration of 1 mg of lipids per 1 ml of chloroform. The needles were left to dry for 1 h and then submerged in a custom electroformation chamber (Fig. 5.2b) filled with 3 ml of 200 mM/ml sucrose solution, with addition of blue food dye. A signal generator was then connected to the needles and the electroformation took place in two steps: in the first one, 10 Hz signal with 5 V (peak-to-peak) was applied for 2 h, then the frequency was lowered to 5 Hz (also at 5 Vpp) for an additional half an hour. Afterwards, the GUVs-rich sucrose solution was transferred to a vial filled with PBS in a proportion of 1:1. Note, that to transfer the GUVs, the tip of the pipette was cut-off to increase the diameter of its orifice and thus reduce the shear stress on GUVs. The suspension was then gently

Fig. 5.1 An image of the PMMA coupon covered with regular grid for measuring displacements using the grid method

Fig. 5.2 Device for testing and producing GUVs. (**a**) A modified PMMA coupon, (**b**) An electroformation chamber connected to a signal generator

(a)

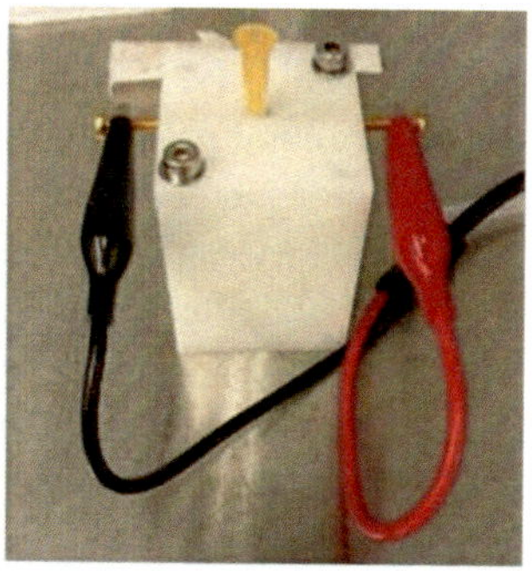

(b)

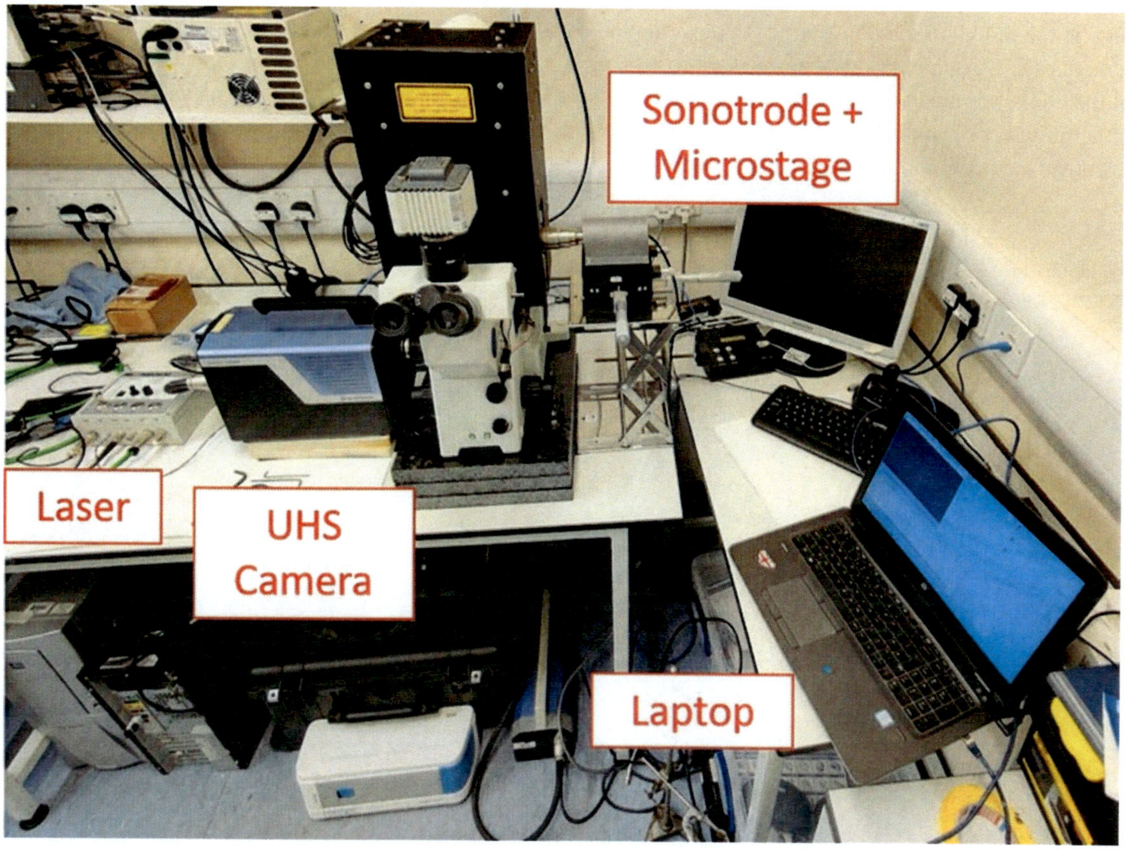

Fig. 5.3 The modified microscope for imaging GUVs using the ultra-high speed camera

mixed and small quantity (250 μl) was transferred into the well on the surface of PMMA. Due of the density difference between the sucrose solution and PBS, the GUVs sink to the bottom of the well and form avidin-biotin link during an incubation period of about 30 min. After the incubation period most of the liquid in the well was gently removed with a pipette, leaving only attached GUVs and a thin film on the surface.

The coupon containing attached GUVs was placed under an inverted microscope (Olympus IX71) that was customized to image the GUVs under ultrasonic deformation (Fig. 5.3). The Shimadzu HPV-X camera was connected to the camera port and a 50× objective was used to record the movement of the GUVs at 500 kfps. Under such magnification one pixel corresponds to 0.64 μm, which was validated using a calibration grid with a pitch of 30 μm. Due to the combination of very high magnification and frame rate the lighting was particularly challenging. Here, we used a pulsed laser (Cavilux, Cavitar) that sends light pulses of up to 150 ns long synchronized with the camera acquisition rate. The coupon was then excited with the sonotrode at 30% of its maximal power of the sonotrode and the GUVs were imaged during the steady-state oscillation.

5.3 Results

During the calibraiton step, the displacement fields were measured using the grid method, spatially smoothed using Gaussian filter with the window size of 35 pixels and temporarily (also using Gaussian filter) with the window size of 5 frames. The displacements were then differentiated spatially using central finite difference and resulting strains were averaged for each cross section and fitted with a sine function to measure frequency and amplitude of the excitation. It was found that the measured frequency (19.8 kHz) was matching the sonotrode's reference 20 kHz, with the small difference being explained by the frequency tracking system of the sonotrode to maximise the resonance. At the chosen level of excitation (30% of maximal power of the sonotrode) the longitudinal strain at the region corresponding to the O-ring ranges between 800 and 1000 μm/m (Fig. 5.4).

A single GUV was isolated and imaged at 500 kfps (Fig. 5.5). A subpixel edge detection algorithm [12] was employed to track the movement of the vesicle and a circle was fitted at each frame to the processed data. The fitting allows to follow both the rigid body displacement (due to downstream effects, misorientation of the sonotrode with the camera frame, etc.) as well as the change in the radius which corresponds to deformation of the GUV. Due to the relatively large magnitude of the rigid body movement (a few pixels) the signal-to-noise ratio was excellent, as shown in Fig. 5.6a. Conversely, the detected change in radius of the GUV was much more subtle with the amplitude of 0.009 pixel (deviation from the mean), which corresponds to the strain amplitude of 886 μm/m (Fig. 5.6b). In both cases the measured frequency was close to 20.0 kHz (20.0 kHz and 19.7 kHz for displacements and the radius respectively).

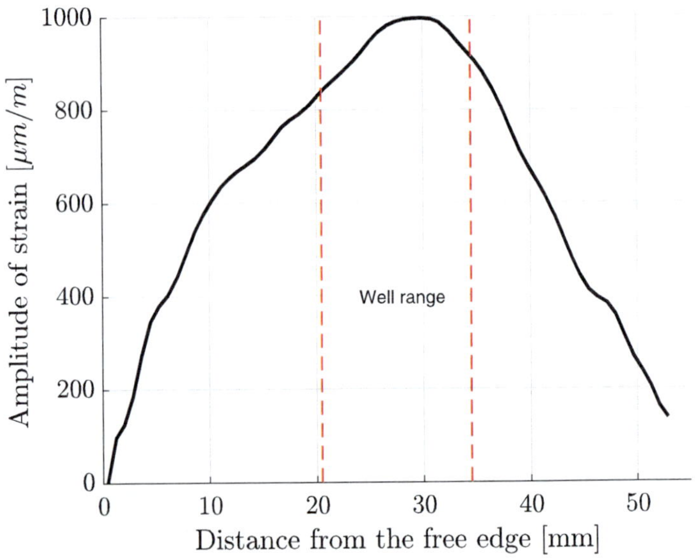

Fig. 5.4 Amplitude of average strain for each cross section along the length of the PMMA coupon

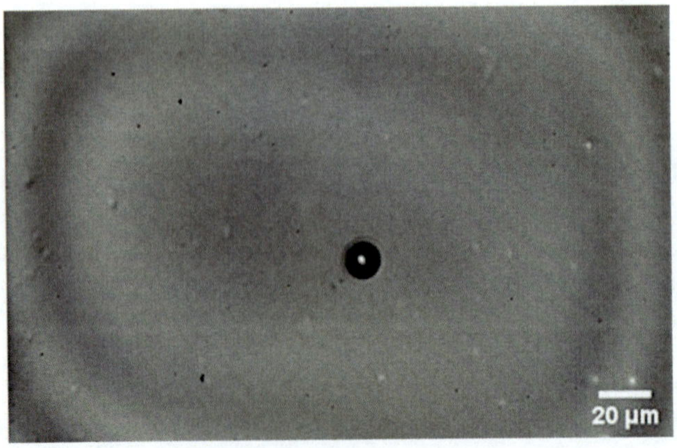

Fig. 5.5 An isolated GUV used to validate the test

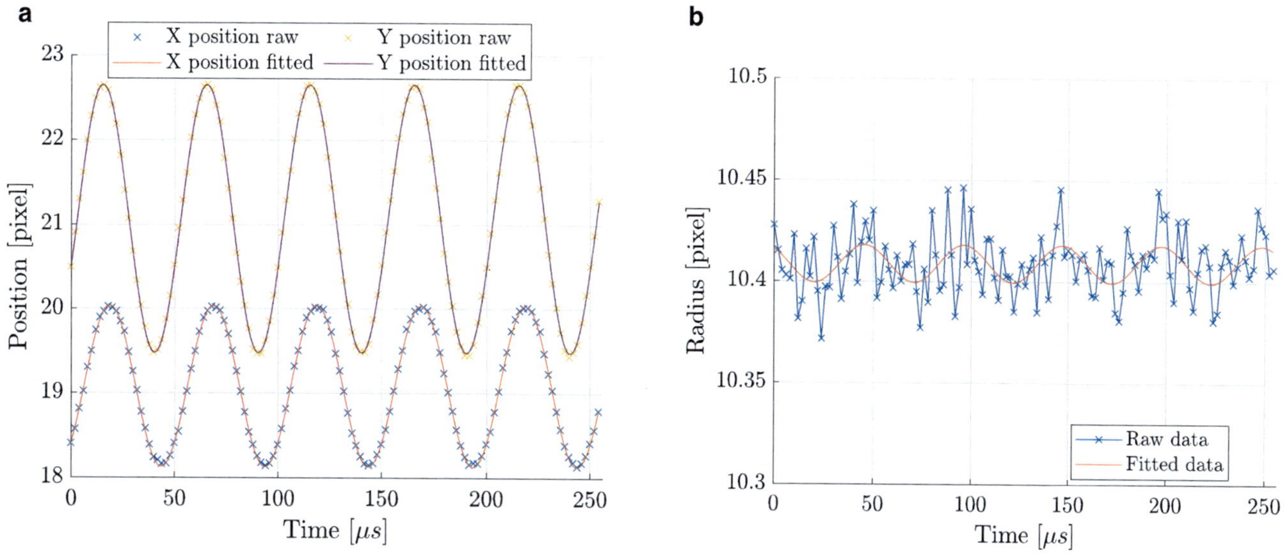

Fig. 5.6 Analysis of the movement and deformation of the imaged GUV. (**a**) Displacement of the centre, (**b**) Evolution of radius (deformation)

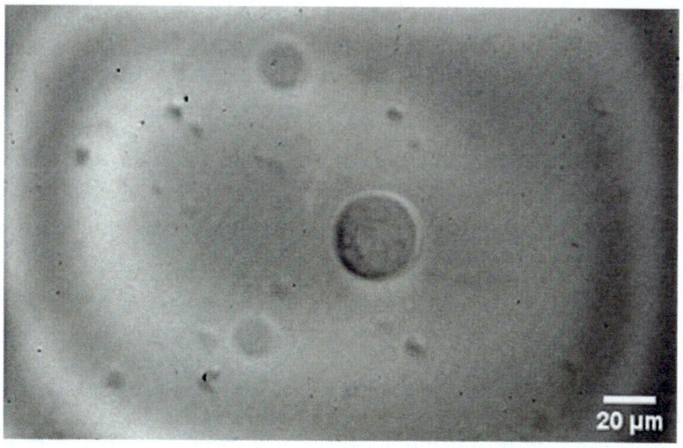

Fig. 5.7 A larger GUV imaged with the ultra-high speed camera with much poorer contrast

5.4 Discussion

The results indicate that the amplitude of the radius change is in good agreement with what was predicted using the macroscopic calibration protocol. This suggests that the proposed protocol can be used to controllably load attached entities by just controlling the amplitude of the sonotrode oscillation, leading to different levels of deformation of the substrate (in the millistrain range). To increase the number of replicates, more wells can be placed on the surface of the substrate, each one being characterized by different amplitude of the strain (e.g. a well centered 10 mm away from the free edge would be expected to deform GUV up to 600 μm/m, see Fig. 5.4).

It is worth noting that analyzed GUV was very small (diameter of roughly 13 μm), thus the strain measurements were very noisy. It had also a very good contrast with the background, likely due to the blue food dye that absorbed the monochromatic (red) laser light. The contrast is critical for subpixel edge detection. In other cases (Fig. 5.7) the algorithm failed to detect the deformation. As the noise floor for the algorithm is constant (~0.01 px) it is possible to further improve the measurements by increasing the size of GUVs in terms of pixels. This can be done in two independent ways: by producing larger GUVs and using higher magnification. The former can be achieved by placing larger volumes of the GUVs suspension in the well (statistically there will be larger GUVs), as well as optimizing the electroformation protocol. The larger magnification can be achieved by either employing more powerful objectives, or by adding an optical element to the microscope (e.g. Olympus IX71 can add additional 1.6× magnification to the current setup).

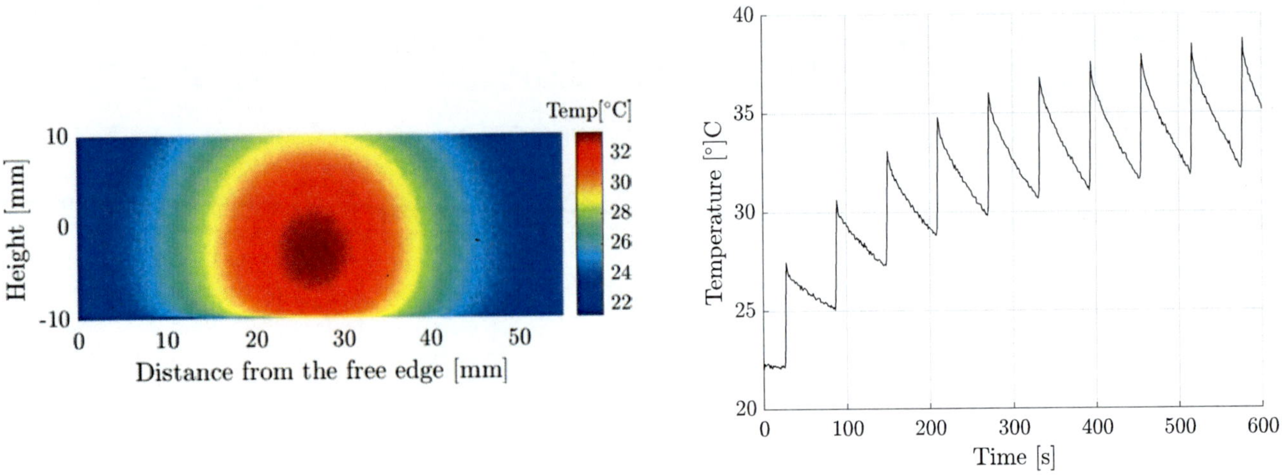

Fig. 5.8 Full-field map of the temperature field. (**a**) Plot of the entire surface of the PMMA coupon, (**b**) History of a single point in the middle of the coupon

Due to high number of cycles per second (20 kHz) the protocol is very promising for studying fatigue of cell membranes, as it would take only 15 min of continuous exposure to reach as much as 20 million cycles. One of practical challenges is the self-heating of the substrate due to viscous energy dissipation. A preliminary study was run by recording the temperature increase at the centre of the PMMA coupon using a high speed infrared detector (Telops M2k), as shown in Fig. 5.8a. The test protocol consisted of sonotrode turning on for 1 s followed by a resting period (sonotrode is off) of 10 s. It was found that such pattern leads to a significant increase in the temperature of PMMA over a few cycles. By increasing the off portion to 60 s it was possible to stabilize the temperature in the middle between 32 and 40 °C (Fig. 5.8b). In future this could be improved on by either employing a small fan next to the experimental setup, or changing the PMMA substrate to a glass one.

5.5 Conclusion and Future Work

The proposed protocol allows to cyclically load attached GUVs to a known deformation level at ultrasonic frequencies. After calibrating the sonotrode/substrate system externally using an ultra-high speed camera the protocol could be used to test the biological response of GUVs/cells regardless of access to such camera, as the deformation could be inferred from the calibration.

One of the interesting applications of such test is to study fatigue of cell membranes modelled with GUVs. The advantage of such model is that it removes a lot of biological factors, such as membrane repair, cell death etc. Simply by increasing the number of cycles at constant deformation the deterioration of the vesicles could be monitored i.e. in terms of increased diffusivity of the GUV. This could be studied by including a fluorophore in the sucrose solution in which the GUVs are formed. Initially only GUVs would produce fluorescent signal which intensity could be tracked and compared against the background as a function of the number of oscillations.

Another interesting application would be to study cells death under ultrasonic deformation. Cells could be grown either in wells or along the entire substrate, each group exposed to different thermo-mechanical history. By sweeping through different amplitudes and durations of pulses the relationship between the ultrasonic excitation and cell death could be established. This is a very promising prospect and will be pursed in the near future.

Finally, to get a better insight into deformation of GUVs (and cells in the future), instead of using ultra-high speed camera to time-resolve the deformation one could employ a high resolution camera and use a stroboscopic technique (e.g. using the pulsed laser employed in this study) to reconstruct the deformation with much better spatial resolution by using the steady-state assumption. Such assumption is well-grounded and the data presented in this contribution supports its validity. This could be particularly important when looking at the deformation of cells as their shape is irregular and couldn't be simply approximated by a change in radius. Instead, other image processing methods should be employed (e.g. digital image correlation).

Acknowledgements The authors gratefully acknowledge EPSRC for funding through grants EP/L026910/1 and EP/R045291/1.

References

1. Amin, A.K., Huntley, J.S., Bush, P.G., Simpson, A.H., Hall, A.C.: Chondrocyte death in mechanically injured articular cartilage: the influence of extracellular calcium. J. Orthop. Res. **27**(6), 778–784 (2009)
2. Marrelli, M., et al.: Dental pulp stem cell mechanoresponsiveness: effects of mechanical stimuli on dental pulp stem cell behavior. Front. Physiol. **9**, 1685 (2018)
3. Uddin, S.M., et al.: Chondro-protective effects of low intensity pulsed ultrasound. Osteoarthr. Cartil. **24**(11), 1989–1998 (2016)
4. Manaka, S., et al.: Low-intensity pulsed ultrasound-induced ATP increases bone formation via the P2X7 receptor in osteoblast-like MC3T3-E1 cells. FEBS Lett. **589**(3), 310–318 (2015)
5. Kudo, N.: High-speed in situ observation system for sonoporation of cells with size- and position-controlled microbubbles. IEEE Trans. Ultrason. Ferroelectr. Freq. Control. **64**(1), 273–280 (2017)
6. Yuan, F., Yang, C., Zhong, P.: Cell membrane deformation and bioeffects produced by tandem bubble-induced jetting flow. Proc. Natl. Acad. Sci. U. S. A. **112**(51), E7039–E7047 (2015)
7. Gao, Q., Cooper, P.R., Walmsley, A.D., Scheven, B.A.: Role of Piezo channels in ultrasound-stimulated dental stem cells. J. Endod. **43**(7), 1130–1136 (2017)
8. Gao, Q., Walmsley, A.D., Cooper, P.R., Scheven, B.A.: Ultrasound stimulation of different dental stem cell populations: role of mitogen-activated protein kinase signaling. J. Endod. **42**(3), 425–431 (2016)
9. Seghir, R., Pierron, F.: A novel image-based ultrasonic test to map material mechanical properties at high strain-rates. Exp. Mech. **58**(2), 183–206 (2017)
10. Grédiac, M., Sur, F., Blaysat, B.: The grid method for in-plane displacement and strain measurement: a review and analysis. Strain. **52**(3), 205–243 (2016)
11. Pereno, V., et al.: Electroformation of giant unilamellar vesicles on stainless steel electrodes. ACS Omega. **2**(3), 994–1002 (2017)
12. Trujillo-Pino, A., Krissian, K., Alemán-Flores, M., Santana-Cedrés, D.: Accurate subpixel edge location based on partial area effect. Image Vis. Comput. **31**(1), 72–90 (2013)

Check for updates

A Mathematical Model of Nitric Oxide Mechanotransduction in Brain

6

Andrew Tamis and Corina S. Drapaca

Abstract

Nitric Oxide (NO) is a diffusible molecule that is involved in many key signaling processes within the brain, notably the regulation of cerebral blood flow and pressure. NO is produced within neurons, endothelial cells, and red blood cells, but is only activated within the endothelial cells by the shear stress at the blood-endothelium interface. Because of the NO significance to brain functionality, various mathematical models of NO behavior have been proposed in literature. However, most of these models do not thoroughly incorporate the NO production in the endothelium through mechanotransduction. In a recent paper, we proposed a mathematical model to describe the steady-state behavior of NO in the brain that accounts for the shear-induced endothelial NO production and the Poiseuille-like flow of blood. In this paper we expand upon this model by introducing a deformable vascular wall and pulsatile blood flow. The arterial wall is modeled as a Maxwell linear viscoelastic material. Numerical simulations will show the mechanical effects on the spatio-temporal distribution of NO.

Keywords

Cerebral nitric oxide · Mechanotransduction · Maxwell viscoelastic model · Oscillations · Reaction-diffusion

6.1 Introduction

Nitric Oxide (NO) is a small gas molecule that is vital to various bio-chemical processes specific to living organisms. Of most note to this study is its role in regulating cerebral blood flow as a neuro-glial-vascular messenger [1–12]. Cerebral NO is produced by synthesis reactions within neurons [5], vascular endothelium and choroid plexus [6], and red blood cells [8], and by shear-induced mechanotransduction at the interface between the endothelium and blood [13]. Diffusion and some specialized chemical processes contribute to the removal of NO from brain [8].

Because of the critical involvement of NO in brain's functions, much literature has been written on mathematically modeling its spatio-temporal dynamics [7, 14–17]. Most initial boundary value problems are reaction-diffusion equations for the NO concentration where the shear-induced production of endothelial NO is a given boundary condition. These models do not incorporate mechanical properties of cerebral blood flow and vasculature and thus cannot predict NO effects on blood flow [1, 5]. In a recent paper [18], we proposed a mathematical model for the steady-state behavior of NO in the brain that accounts for the shear-induced endothelial NO production and the Poiseuille-like flow of blood. Based on the experimental data in [13], we introduced in the reaction-diffusion equation of NO a production term proportional to the NO concentration. The production rate was taken to be proportional to the viscous dissipation at the endothelium-blood boundary. The vascular

A. Tamis · C. S. Drapaca (✉)
Department of Engineering Science and Mechanics, College of Engineering, Pennsylvania State University, University Park, PA, USA
e-mail: csd12@psu.edu

J. Notbohm et al. (eds.), *Mechanics of Biological Systems and Materials & Micro-and Nanomechanics & Research Applications*, Conference Proceedings of the Society for Experimental Mechanics Series,
https://doi.org/10.1007/978-3-030-59765-8_6

wall was assumed to be rigid and the blood was Newtonian with a Poiseuille flow. In this paper we expand upon this model by introducing a deformable vascular wall and pulsatile blood flow. The arterial wall is modeled as a Maxwell linear viscoelastic material and we use the results in [19] to calculate a new viscous dissipation. Numerical simulations show the effects of the dynamics and mechanical properties of the vascular wall on the spatio-temporal distribution of NO concentration.

The structure of the paper is as follows. A brief review of the results in [19] and our mathematical model are given in the next section, which is followed by a section of results. The last section presents our conclusions and future work.

6.2 Mathematical Model

As in [18], we model the geometric domain as concentric horizontal axial symmetric circular cylinders (Fig.6.1). The region occupied by the blood is the lumen of radius a. The endothelium of thickness h is the site of the shear-induced NO. A region of thickness d made of other layers of the vascular wall and extracellular space separates the endothelium and a group of neurons that produce NO. The thickness of the region filled with the neurons is g.

We assume now that the endothelium is made of a homogeneous, Maxwell linear viscoelastic material that is tethered at its outer boundary and exposed to oscillations due to the heart pulsations at its inner boundary [19]. We denote by (r,θ,x) the cylindrical coordinates. The two terms relevant to our work are $\xi(r,\theta,x,t)$, the axial displacement of the endothelium, and $\sigma_{rx}(r,\theta,x,t)$, the shear stress of the endothelium. The assumption of axial symmetry and the fact that the spatial variations can be neglected because the vessel length is much less than the propagating wavelength (approximately 10 m for a systemic circulation at a frequency of 1 Hz) reduce the equation of motion for ξ to the following expression [19]:

$$\rho\frac{\partial^2\xi}{\partial t^2}=\frac{\partial\sigma_{rx}}{\partial r}+\frac{\sigma_{rx}}{\mathrm{r}} \tag{6.1}$$

with ρ the mass density. Equation (6.1) can be further simplified by neglecting the curvature term $\frac{\sigma_{rx}}{\mathrm{r}}$ which is much less than $\frac{\partial\sigma_{rx}}{\partial r}$ since the endothelium's thickness h is much smaller than the vessel radius a. Then, Eq. (6.1) reduces to [19]:

$$\rho\frac{\partial^2\xi}{\partial t^2}=\frac{\partial\sigma_{rx}}{\partial r} \tag{6.2}$$

The constitutive equation of a Maxwell viscoelastic material is:

$$\frac{\partial\sigma_{rx}}{\partial t}=E\left(\frac{\partial\varepsilon_{rx}}{\partial t}-\frac{\sigma_{rx}}{\mu}\right) \tag{6.3}$$

with E the modulus of elasticity, μ the viscosity of the endothelium, and ε_{rx} the infinitesimal strain.

Since the spatial variations are neglected, the infinitesimal strain is in this case:

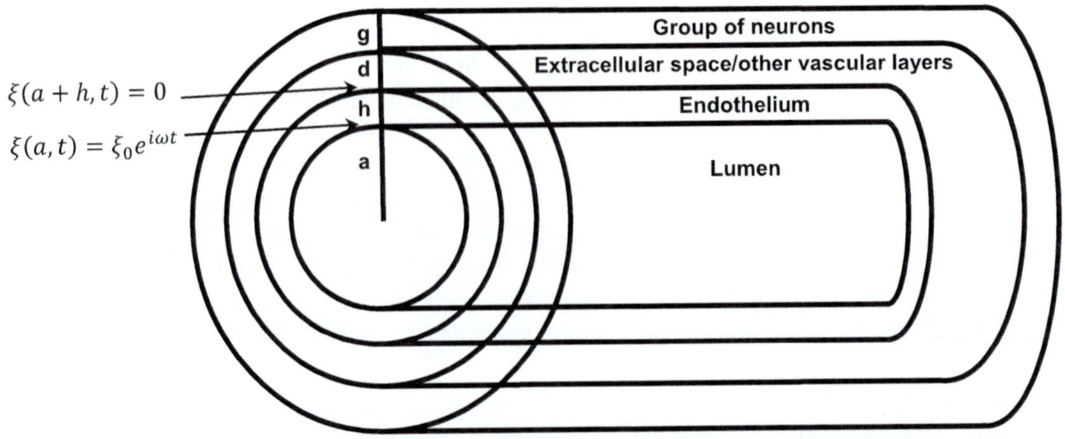

Fig. 6.1 The concentric horizontal axial symmetric circular cylinders represent the lumen of radius a, the endothelium of thickness h, a layer of thickness d filled with other vascular layers and extracellular space, and a layer of thickness g filled with neurons. The endothelium is tethered at $r = a + h$ and is subjected to oscillations of frequency ω at $r = a$. The equations representing these boundary conditions are also shown, where ξ is the axial displacement of the endothelium

$$\varepsilon_{rx} = \frac{\partial \xi}{\partial r} \tag{6.4}$$

Replacing formula (6.4) in Eq. (6.3) and introducing the scalar $\gamma = \frac{\omega\mu}{E}$ give:

$$\frac{\partial \sigma_{rx}}{\partial t} = E\frac{\partial^2 \xi}{\partial t \partial r} - \frac{\omega}{\gamma}\sigma_{rx} \tag{6.5}$$

Lastly, substituting Eq. (6.5) in Eq. (6.2) yields [19]:

$$\rho\frac{\partial^3 \xi}{\partial t^3} - E\frac{\partial^3 \xi}{\partial t \partial r^2} + \frac{\rho\omega}{\gamma}\frac{\partial^2 \xi}{\partial t^2} = 0 \tag{6.6}$$

The endothelium is assumed to be tethered at its outer boundary ($r = a + h$) and exposed to oscillations at its inner boundary ($r = a$). The corresponding mathematical representations are Dirichlet boundary conditions:

$$\xi(a+h,t) = 0,$$

$$\xi(a,t) = \xi_0 e^{i\omega t}$$

The solution to Eq. (6.6) with the above boundary conditions is [19]:

$$\xi(r,t) = \frac{e^{\alpha(h-r+a)} - e^{-\alpha(h-r+a)}}{e^{\alpha h} - e^{-\alpha h}}\xi_0 e^{i[\omega t - \beta(r-a)]} \tag{6.7}$$

with:

$$\alpha^2 = \frac{\rho\omega^2}{2E}\left[\sqrt{1+\frac{1}{\gamma^2}} - 1\right], \beta^2 = \frac{\rho\omega^2}{2E}\left[\sqrt{1+\frac{1}{\gamma^2}} + 1\right]$$

The shear stress is further found by replacing formula (6.7) into Eq. (6.2) and integrating:

$$\sigma_{rx}(r,t) = \frac{\xi_0 E}{\sqrt{1+\frac{1}{\gamma^2}}}\frac{(\alpha - i\beta)e^{\alpha(h-r+a)} + (\alpha + i\beta)e^{-\alpha(h-r+a)}}{e^{\alpha h} - e^{-\alpha h}}e^{i[\omega t - \beta(r-a)]} \tag{6.8}$$

The highest value of the shear stress is at the endothelium-lumen interface [19]:

$$\sigma_{rx}(a,t) = \tau_W = \frac{\xi_0 E}{\sqrt{1+\frac{1}{\gamma^2}}}\frac{(\alpha - i\beta)e^{\alpha h} + (\alpha + i\beta)e^{-\alpha h}}{e^{\alpha h} - e^{-\alpha h}}e^{i\omega t} \tag{6.9}$$

Lastly, replacing formula (6.7) into formula (6.4) gives:

$$\varepsilon_{rx} = \frac{\partial \xi}{\partial r} = \frac{\xi_0 e^{i[\omega t - \beta(r-a)]}}{e^{\alpha h} - e^{-\alpha h}}\left[-(\alpha + i\beta)e^{\alpha(h-r+a)} - (\alpha - i\beta)e^{-\alpha(h-r+a)}\right]$$

and thus:

$$\frac{\partial \varepsilon_{rx}}{\partial t} = \frac{\xi_0 i\omega}{e^{\alpha h} - e^{-\alpha h}}\left[-(\alpha + i\beta)e^{\alpha(h-r+a)} - (\alpha - i\beta)e^{-\alpha(h-r+a)}\right]e^{i[\omega t - \beta(r-a)]} \tag{6.10}$$

We can introduce now the viscous dissipation as:

$$\frac{\sigma_{rx}}{\tau_W}\frac{\partial \varepsilon_{rx}}{\partial t} = \frac{-i\xi_0\omega(\alpha^2 + \beta^2)\left[e^{2\alpha(h-r+a)} + e^{-2\alpha(h-r+a)} + 2\right]}{(\alpha - i\beta)e^{2\alpha h} - (\alpha + i\beta)e^{-2\alpha h} + 2i\beta}e^{i[\omega t - 2\beta(r-a)]} \tag{6.11}$$

As in [18], we assume that NO is produced in the region $[a, a+h] \cup [a+h+d, a+h+d+g]$ and is removed in the region $[0, a] \cup [a+h, a+h+d+g]$. The radial diffusion of NO happens throughout the entire domain $[0, a+h+d+g]$. The balance law of mass in cylindrical coordinates for C_{NO}, the concentration of NO, is:

$$\frac{\partial C_{NO}}{\partial t} = D_{NO}\left(\frac{\partial^2 C_{NO}}{\partial r^2} + \frac{1}{r}\frac{\partial C_{NO}}{\partial r}\right) + v_1 H\left(r-(a+h+d)\right) - \frac{v_{\max} C_{NO}}{K_{\max} + C_{NO}} H\left(r-(a+h)\right) + \frac{\sigma_{rx}}{\tau_W}\frac{\partial \varepsilon_{rx}}{\partial t}\left[H(r-a) - H\left(r-(a+h)\right)\right]C_{NO} - \lambda\left(1-H(r-a)\right)C_{NO} \tag{6.12}$$

In Eq. (6.12), D_{NO} is the diffusion coefficient for NO, v_1 is the constant rate of NO synthesis by neurons, v_{max} is the maximum rate at saturating concentration in the region $[a+h, a+h+d+g]$, and K_{max} is the NO concentration at which the reaction rate is $\frac{v_{\max}}{2}$ in the same region. The rate of NO decay due to blood's hemoglobin is denoted by λ and is assumed constant. The fourth term of the right-hand side of Eq. (6.12) models the shear-induced mechanotransduction of NO in the endothelium where the viscous dissipation is given by formula (6.11). Lastly, we denoted by H the Heaviside step function.

Equation (6.12) is solved numerically for zero initial conditions and the following mixed boundary conditions [18]:

$$\frac{\partial C_{NO}}{\partial r}(0,t) = 0, C_{NO}(a+h+d+g,t) = 10^{-8}\left[\frac{mol}{m^3}\right] \tag{6.13}$$

The built-in function **pdepe** in Matlab [20] was used to solve the initial boundary value problem (6.12) and 6.13). The function combines a variable-step, variable-order numerical scheme for time discretization and a finite element solver for space discretization.

6.3 Results

The parameters used in the numerical simulations are given in Table 6.1.

The two parameters that affect the spatio-temporal distribution of NO concentration are the amplitude of oscillations ξ_0 and the mechanical parameter γ. The results of a sensitivity analysis performed on these two parameters are shown in Fig. 6.2. For a fixed $\gamma = 0.25$, the production of endothelial NO increases with increasing amplitude of oscillations (Fig. 6.2a for $r = 25.5 \times 10^{-6} m$), while for a fixed $\xi_0 = .6 x\, 10^{-3}\, m$ the production of endothelial NO increases with increasing γ from a more viscoelastic endothelium ($\gamma = 0.25$) to a more elastic one ($\gamma = 0.75$) (Fig. 6.2b for $r = 25.5 \times 10^{-6} m$). The healthy regime for γ is $\gamma = 0.1 \div 0.25$ [19], but a range for healthy amplitudes of oscillations ξ_0 is not known. It is possible that the values we used in our simulations for ξ_0 are too big (order $O(10^{-3})$) and thus the production of endothelial NO due to shear stress might not be so pronounced. Here we modeled the effect of blood flow as an oscillatory boundary condition; however, a coupled mechanical model of the pulsatile blood flow and

Table 6.1 List of parameters with corresponding values and units

Considerations	Parameters	Values and units [Reference]
Geometry	a	$25 \times 10^{-6} m$ [15]
	h	$0.5 \times 10^{-6} m$ [15]
	d	$4 \times 10^{-6} m$
	g	$5 \times 10^{-6} m$ [7]
NO diffusion, synthesis & inactivation	D_{NO}	$3.3 \times 10^{-9} m^2/s$ [7, 15]
	v_1	$1.6 \times 10^{-3} mol/(m^3\, x\, s)$ [7]
	v_{max}	$2 \times 10^{-3} mol/(m^3\, x\, s)$ [7]
	K_{max}	$10^{-6} mol/m^3$ [7]
	λ	$2.3 \times 10^2 \frac{1}{s}$ [16]
Maxwell viscoelastic endothelium	ρ	$1\, Kg/m^3$
	E	$6 \times 10^3\, N/m$ [19]
	ω	$1\, Hz$ [19]
	ξ_0	$0.4 \div 1 \times 10^{-3}\, m$
	γ	$0.25 \div 0.75$

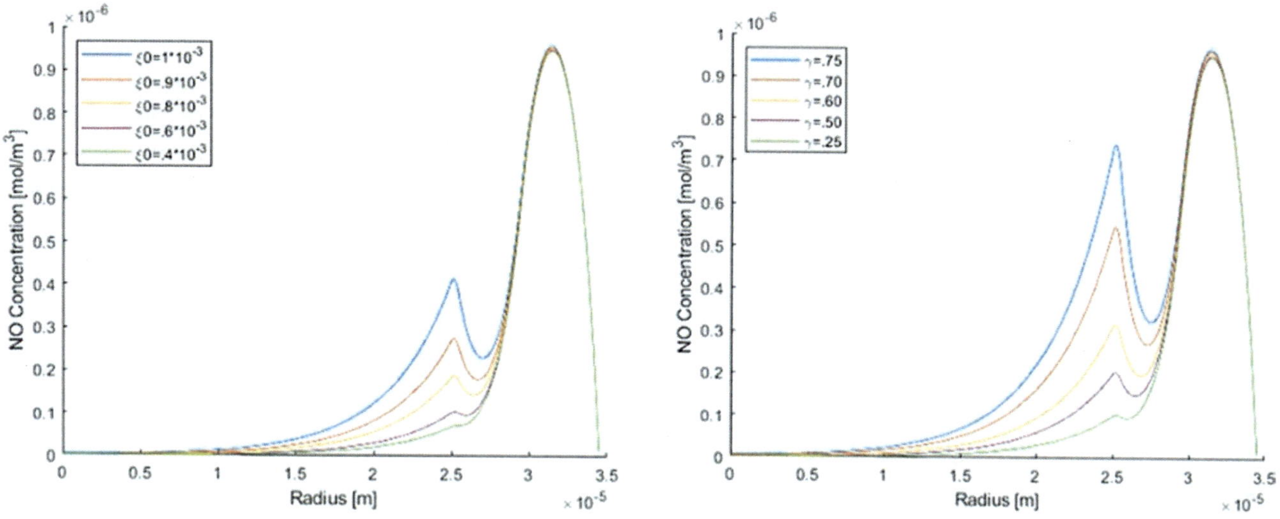

(a) (b)

Fig. 6.2 Plots of NO concentration for (**a**) various values of ξ_0 at fixed $\gamma = 0.25$, and (**b**) various values of γ at fixed $\xi_0 = 0.6 \times 10^{-3}\ m$

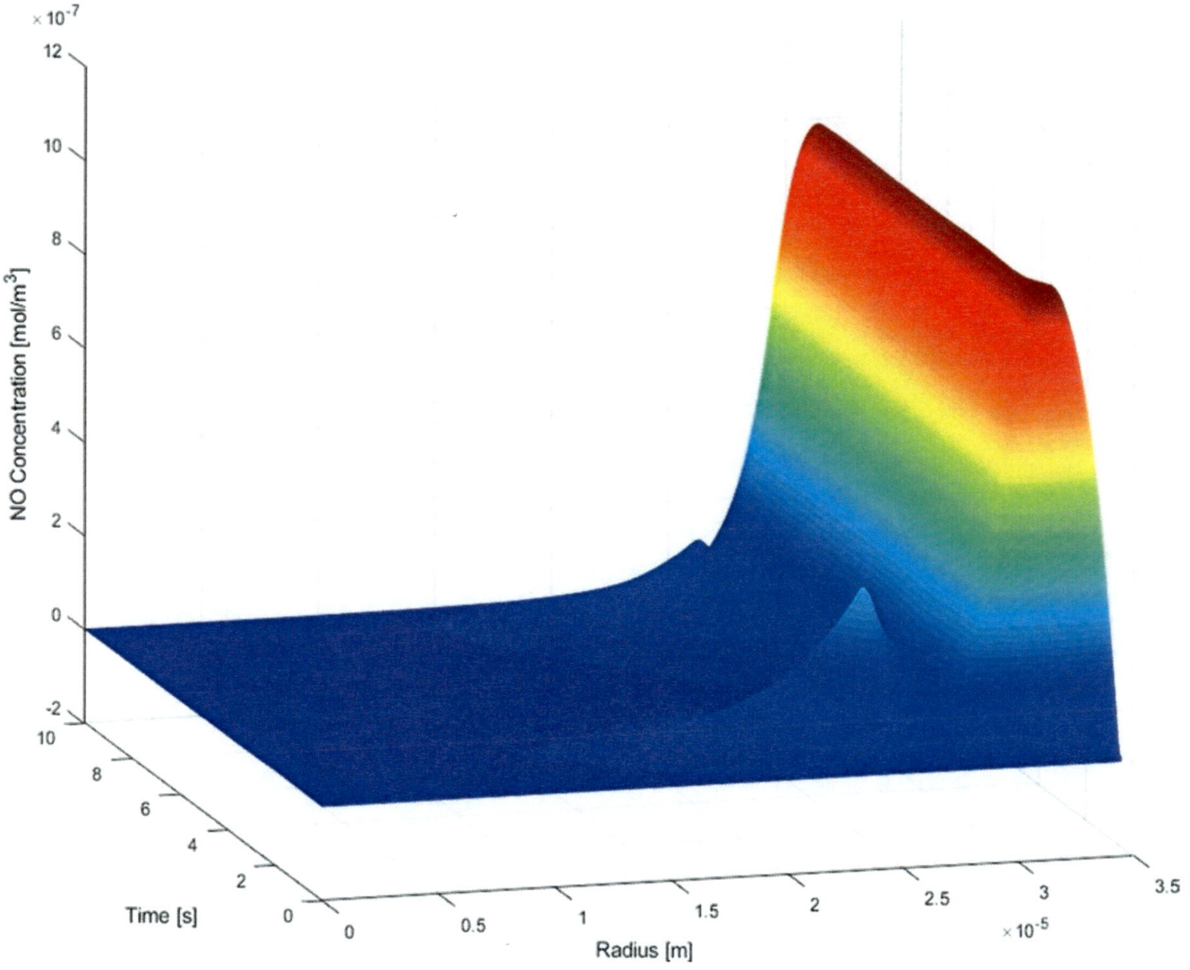

Fig. 6.3 Spatio-temporal plot of NO concentration for $\gamma = 0.25$, $\xi_0 = 1 \times 10^{-3}\ m$

endothelium dynamics could suggest more realistic values for ξ_0. Nevertheless, the proposed model shows that the shear-induced production of endothelial NO depends on the mechanical properties of the endothelium and on the frequency and amplitude of the pulsatile blood flow. Lastly, a three-dimensional plot of the spatio-temporal distribution of NO concentration is shown in Fig. 6.3. The oscillatory behavior at endothelium is apparent.

6.4 Conclusion

In this paper we proposed a one-dimensional, reaction-diffusion equation for the concentration of NO that includes a new production term for the shear-induced mechanotransduction happening at the endothelium-lumen interface. This term was assumed to be proportional to the NO concentration and the production rate was proportional to the viscous dissipation in the endothelium. The endothelium was deformable and made of a Maxwell linear viscoelastic material. We used the results in [19] to calculate the viscous dissipation. The blood flow was modelled as an oscillatory boundary condition on the inner layer of the endothelium. Our numerical simulations show that the production of endothelial NO depends on the amplitude and frequency of the pulsatile blood flow and the mechanical properties (stiffness and viscosity) of the endothelium. In our future work we plan to couple this model to a mechanical model of the pulsatile blood flow. Such a mathematical model could contribute to the development of successful NO-based therapies for brain diseases associated with disturbances in NO production and/or signaling.

References

1. Attwell, D., Buchan, A., Charpak, S., Lauritzen, M., MacVicar, B.A., Newman, E.A.: Glial and neuronal control of brain blood flow. Nature. **468**, 232–243 (2010). https://doi.org/10.1038/nature09613
2. Barbosa, R.M., Lourenco, C.F., Santos, R.M., Pomerleau, F., Huettl, P., Gehardt, G.A., Laranjinha, J.: In vivo real-time measurement of nitric oxide in anesthetized rat brain. Methods Enzymol. **441**, 351–367 (2008). https://doi.org/10.1016/S0076-6879(08)01220-2
3. Buerk, D.G., Ances, B.M., Greenberg, J.H., Detre, J.A.: Temporal dynamics of brain tissue nitric oxide during functional forepaw stimulation in rats. Neuroimage. **18**, 1–9 (2003)
4. Contestabile, A., Monti, B., Polazzi, E.: Neuronal-glial interactions define the role of nitric oxide in neural functional processes. Curr. Neuropharmacol. **10**(4), 303–310 (2012). https://doi.org/10.2174/157015912804143522
5. Forstermann, U., Sessa, W.C.: Nitric oxide synthases: regulation and function. Eur. Heart J. **33**, 829–837 (2012). https://doi.org/10.1093/eurheartj/ehr304
6. Garry, P.S., Ezra, M., Rowland, M.J., Westbrook, J., Pattinson, K.T.S.: The role of the nitric oxide pathway in brain injury and its treatment – from bench to bedside. Exp. Neurol. **263**, 235–243 (2015). https://doi.org/10.1016/j.expneurol.2014.10.017
7. Hall, C.N., Garthwaite, J.: Inactivation of nitric oxide by rat cerebellar slices. J. Physiol. **577**(2), 549–567 (2006). https://doi.org/10.1113/jphysiol.2006.118380
8. Helms, C.C., Liu, X., Kim-Shapiro, D.B.: Recent insights into nitrite signaling processes in blood. Biol. Chem. **3**, 319–329 (2016). https://doi.org/10.1515/hsz-2016-0263
9. Ledo, A., Barbosa, R.M., Gerhardt, G.A., Cadenas, E., Laranjinha, J.: Concentration dynamics of nitric oxide in rat hippocampal subregions evoked by stimulation of the NMDA glutamate receptor. Proc. Natl. Acad. Sci. U. S. A. **102**(48), 17483–17488 (2005). https://doi.org/10.1073/pnas.0503624102
10. Mishra, A.: Binaural blood flow control by astrocytes: listening to synapses and the vasculature. J. Physiol. **595**(6), 1885–1902 (2017). https://doi.org/10.1113/JP270979
11. Moncada, S., Palmer, R.M.J., Higgs, E.A.: Nitric oxide: physiology, pathophysiology, and pharmacology. Pharmacol. Rev. **43**(2), 109–142 (1991)
12. Santos, R.M., Lourenco, C.F., Ledo, A., Barbosa, R.M., Laranjinha, J.: Nitric oxide inactivation mechanisms in the brain: role in bioenergetics and neurodegeneration. Int. J. Cell Biol. **2012**, 391914 (2012). https://doi.org/10.1155/2012/391914
13. Sriram, K., Laughlin, J.G., Rangamani, P., Tartakovsky, D.M.: Shear-induced nitric ox- ide production by endothelial cells. Biophys. J. **111**, 208–221 (2016). https://doi.org/10.1016/j.bpj.2016.05.034
14. Buerk, D.G.: Can we model nitric oxide biotransport? A survey of mathematical models for a simple diatomic molecule with surprisingly complex biological activities. Annu. Rev. Biomed. Eng. **3**, 109–143 (2001). https://doi.org/10.1146/annurev.bioeng.3.1.109
15. Kavdia, M., Tsoukias, N.M., Popel, A.S.: Model of nitric oxide diffusion in an arteriole: impact of hemoglobin-based blood substitute. Am. J. Physiol. Heart Circ. Physiol. **282**, H2245–H2253 (2002). https://doi.org/10.1152/ajpheart.00972.2001
16. Vaughn, M.W., Kuo, L., Liao, J.C.: Effective diffusion distance of nitric oxide in the microcirculation. Vasc. Physiol. **274**(5), H1705–H1714 (1998). https://doi.org/10.1152/ajpheart.1998.274.5.H1705
17. Vaughn, M.W., Kuo, L., Liao, J.C.: Estimation of nitric oxide production and reaction rates in tissue by use of a mathematical model. Am. J. Physiol. **274**(6), H2163–H2176 (1998). https://doi.org/10.1152/ajpheart.1998.274.6.H2163
18. Drapaca, C.S., Tamis, A.: Mathematical modeling of the steady-state behavior of nitric oxide in brain *Proceedings of AMMCS 2019* (under review)
19. Hodis, S., Zamir, M.: Solutions of the Maxwell viscoelastic equations for displacement and stress distributions within the arterial wall. Phys. Rev. E. **78**, 021914 (2008)
20. MATLAB R2019a, The MathWorks, Inc., Natick, Massachusetts, United States

Check for updates

Evaluating the Application of DIC on Heartbeat Detection by Using a Self-developed Artery Vessel Simulator

7

Chi-Hung Hwang, Yu-Chieh Chen, Rui-Cian Weng, Wei-Chung Wang, Tzu-Yu Kuo, Ta-Wei Kan, and Chih-Hao Chang

Abstract

In this study, a self-developed artery vessel simulator is made to simulate an artery vessel inside an organ by casting a PDMS hollow-cylinder with a 4-mm through-depth hole. Intermittently pumping water into the hollow-cylinder to simulate the blood flow through the vessel. Water flows through the through-depth hole and interacts with the cast PDMS structure to introduce varied pressure on the inner surface and caused the PDMS structure vibrating, the associated deformation and strain are then analyzed by adopting the stereo-digital image correlation method. The pumping behavior is verified by comparing with the radial artery vessel mechanical behaviors of a young volunteer objective, the comparison indicates the proposed artery simulator can be considered as a usable design according to current performance. In addition, the study also concludes that the artery vessels might be more easily identified by using strain filed rather than the displacement filed.

Keywords

Digital image correlation method , Artery vessel , PDMS , ECG , Intermittently pumping

7.1 Introduction

This study is part of a project which is focusing on the development of a fluorescent imaging system for general surgery, particularly for detecting the possible vessel leakage and the mechanical behaviors of the organ after suture. Attaching sensors on the surface of an organ might introduce extra damage to the organ, and is not always acceptable by the patients and surgical doctors. Therefore, the digital image correlation method (DIC), an image-based deformation/displacement measurement method, is adopted as an evaluation method in this ongoing project. The DIC has been widely used to determine the mechanical behaviors in different fields by tracking the natural or artificial random pattern on the object surfaces recorded as images which are captured at different status. Creating the artificial surface random pattern is necessary while adopting the DIC method for measurement, in general, cannot avoid contact with the object surface and might introduce extra contact force to the object. However, the DIC method is always considered as a non-contact measurement method because the currently available artificial pattern generating methods always introduce minor and ignorable forces to the surface. Thanks to non-contact property, the DIC method extends its applications on determining the mechanical properties of the soft-materials and biomaterials. There are various applications of the DIC methods on the bio-materials during the last two decades, such

C.-H. Hwang (✉) · Y.-C. Chen · R.-C. Weng
National Applied Research Laboratories, Taiwan Instrument Research Institute, Hsinchu, Taiwan
e-mail: chhwang@narlabs.org.tw

W.-C. Wang · T.-Y. Kuo · T.-W. Kan · C.-H. Chang
Department of Power Mechanical Engineering, National Tsing Hua University, Hsinchu, Taiwan

J. Notbohm et al. (eds.), *Mechanics of Biological Systems and Materials & Micro-and Nanomechanics & Research Applications*, Conference Proceedings of the Society for Experimental Mechanics Series,
https://doi.org/10.1007/978-3-030-59765-8_7

as determining the mechanical properties of cells by using fluorescent particles as random tracking marks [1], determining the mechanical properties of muscle and bone [2, 3], and determining the ventricular volume-change when the patient subjected to the cardiac surgery [4–6].

Directly access animal or human organs for evaluating the developed imaging system is not allowed, unless the performance evaluation of the imaging system and the system configuration is well developed. The purpose of the developed simulator is to provide a test platform for the development of the fluorescent imaging system to minimize the possible risk for the fluorescent imaging system development. By integrating a DC membrane pump with a signal amplifier with a pumping driving circuit, a PDMS cast structure, and an ECG signal generator, the assembled artery vessel simulator can mimic the local organ surface mechanical behavior introduced by the blood flows through an organ.

Intermittent pumping mode associated with the generated ECG signal drives water to flow through the simplified artery-organ structure. A stereo-camera pair installed above the simulator takes images as water intermittently pumped through the artery-organ structure at different status. Then a commercial available stereo-DIC software analyzes the captured images to determine the displacement and strain field around the 'artificial artery.' The developed simulator with the proposed intermittent water pumping mode is verified by comparing the strain field obtained from the 'artificial artery' of the developed simulator, and the one above the radial-artery vessel of a young volunteer objective.

7.2 The Artery Simulator

The simulator, as shown in Fig. 7.1, using a computer to generate the ECG signal, store the stereo-images taken from the simulator, and to determine the deformation/strain field. Figure 7.2 is the typical ECG signal generated with a maximum of 1 voltage in amplitude. A DAQ and a home-made pump driver integrated with a power supply amplifies the generated signal up to the pump operation voltage level and drives the DC membrane pump to operate intermittently. The DC membrane pump driven by the ECG signals intermittently draws water from a tank and delivers the water flow through the connecting pipes, the PDMS structure, and then flows back to the water tank to complete the liquid circulation.

The PDMS structure, as shown in Fig. 7.3, the core of the simulator is cast by filling the cast-model with pre-mixed pre-polymer and the cross-linker compounds at a 13.8 weight ratio. To simulate an artery below the shallow surface of an organ, a hollow cylinder cast from PDMS with 100 mm in height with the inner and outer diameter is Ø50 mm and Ø70 mm, respectively. The PDMS structure possesses a Ø4 mm through-depth hole locates 3 mm below the outer surface to simulate an artery vessel below the surface of an organ. Inserting a 120 mm straight metal rod into the model while casting the PDMS, after curing, a Ø4 mm through-depth hole uses to simulate the artery below the skin is prepared. For discussion convenience,

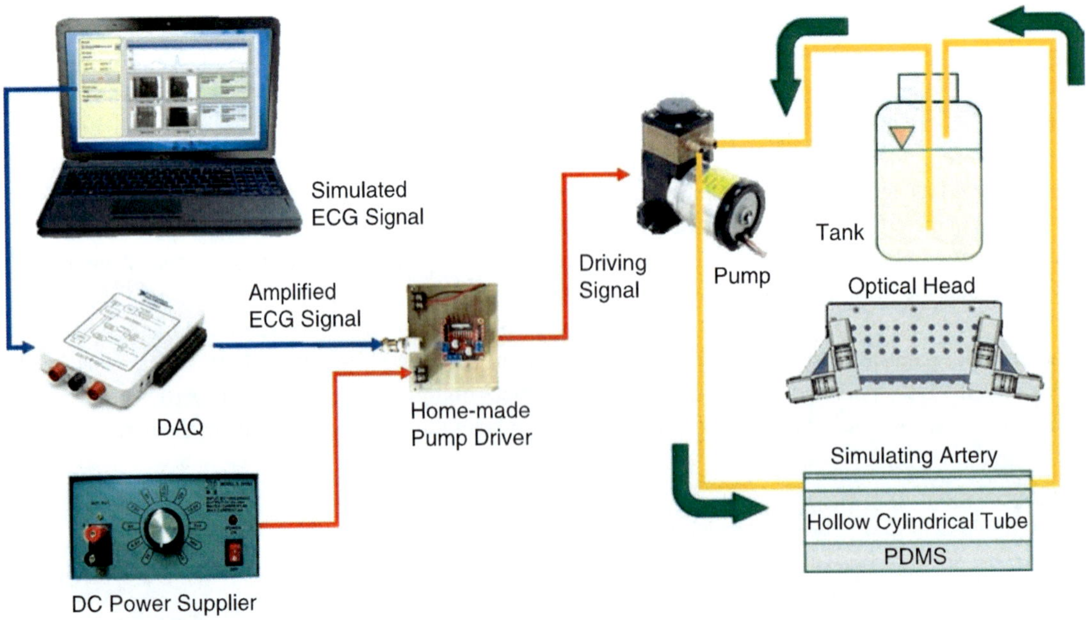

Fig. 7.1 An artery simulator

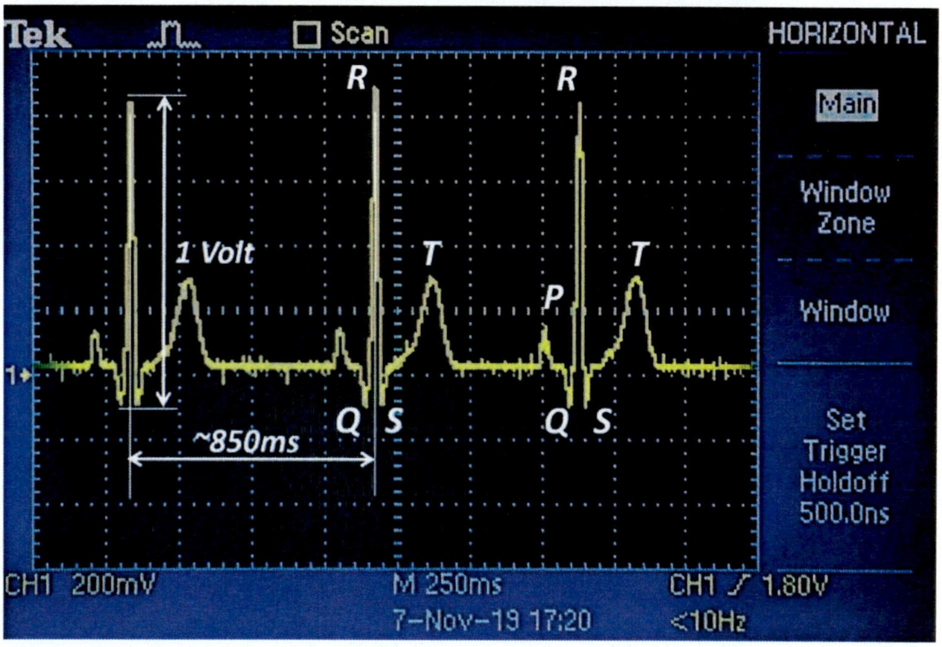

Fig. 7.2 Generated ECG signal

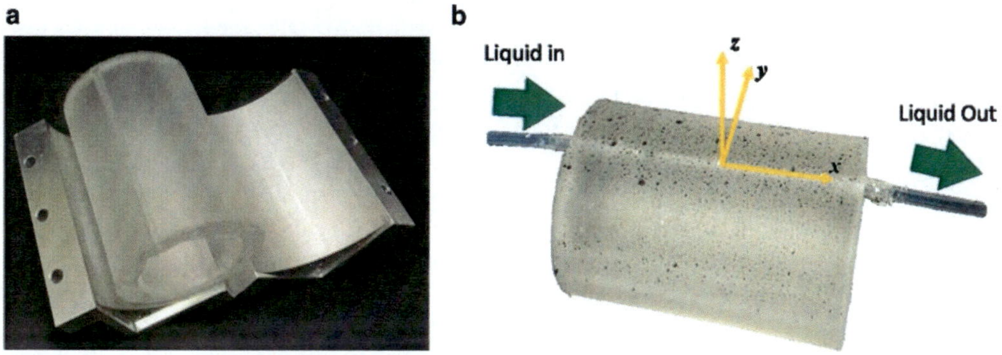

Fig. 7.3 The casted PDMS hollow cylinder with through depth hole is made to simulate artery-organ structure. (**a**) Departing the casted PDMS from a part of casting model. (**b**) Connecting plastic tubes at the both ends of through hole of the casted PDMS hollow cylinder

the coordinate of the PDMS structure is given in Fig. 7.3b. The DC membrane pump is driving by the generated ECG signal to pump water into a 4 mm "artificial vessel" to model the artery with a 400 ml/min maximum flow rate. A camera-pair uses to take images for calculating the stereo-DIC displacement/strain field. The ECG signal uses for driving the pump also triggers the cameras for imaging; different from pump driving, cameras are only triggered when the R-peak detected. In this developed system, the corresponding time delay between the R-peak detected and the camera triggered and captured is less than 15 μs.

For comparison purposes, one of the authors volunteers to be the test objective for taking a series of wrist images to obtain the skin strain field above the radial artery, as shown in Fig. 7.4. The skin above the radial artery is rendered with black ink to create the artificial random pattern for future DIC analysis use.

7.3 Results and Discussions

To perform a test, the computer first sends out the generated ECG signal to the self-developed pump driver and the DAQ at the same time. The self-developed pump driver then amplifies the generated signal to control power on/off for the DC membrane pump for pumping water to circulate along the tubes and PDMS structure. At the same time, the DAQ digitalized the received ECG signal and send it to the computer for R-peak detection; whenever the R-wave peak detected, the computer sends out a triggering signal to activate the cameras for taking images.

Fig. 7.4 The skin around the radial artery is rendered with black ink to generate traceable pattern

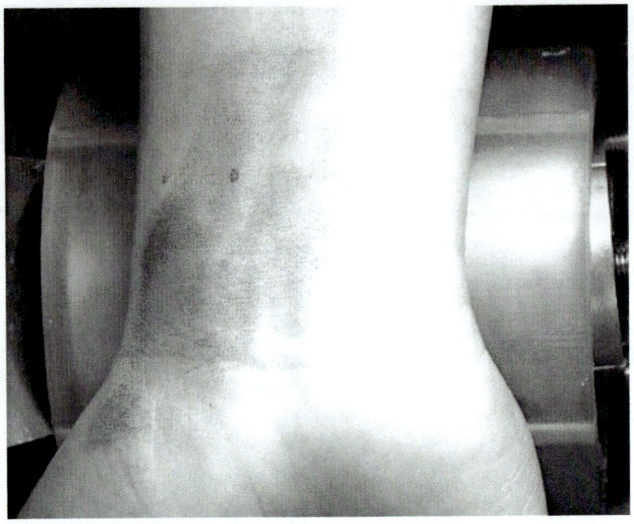

From the captured images, the U, V, and W displacement at the center of the artery-organ PDMS structure is calculated and illustrated in Fig. 7.6. The results show that the U and W displacement are related small in comparison with the V displacement. However, it is interesting to know that the V displacement is continuously increased for the first 130 frames and become stable after. As for the magnitude of the displacement, U and V displacement are always below 0.1 mm, and V-displacement is bounded by 0.44 mm. Analyzing the captured images with 3D-DIC, the U-, V-, and W- displacement at the center of the artery-organ PDMS structure is determined and illustrated in Fig. 7.6. The results show that the U- and W-displacements are related small in comparison with the V- displacement. Interestingly, the V- displacement is continuously increased for the first 130 frames and become stable after then. As for the magnitude of the displacement, U-, W- displacement are both below 0.1 mm, and the V-displacement is bounded to 0.44 mm. In this case, at the selected point right above the artificial artery (the through-depth hole along with the PDMS structure), V displacement is about four times higher to W displacement, eight times of U displacement. It is worth to notice that the PDMS structure is not fixed and will vibrate while water is intermittently pumping into the system; considering the geometrical configuration of the PDMS structure, the stiffness along x-direction should be the highest concerning the other ones along y- and z-directions; however, based on the displacement results, it is interesting to know the displacement along y-direction is the higher than the one along z-direction that indicates the stiffness of z-direction is higher than y-direction for this cast PDMS structure. According to the displacement plot shown in Fig. 7.5, three displacements of a defined point become stable after several frames counted; for example, the V-displacement is continuously increasing before the 130th frame counted; the U-displacement and W-displacement are diverging before the 50th and the 100th frame, respectively. In this study, the stereo-imaging camera pair captures images after the ECG R-peak event detected.

Figure 7.6 shows the ε_{xx}, ε_{yy} and ε_{xy} strain maps in four sequential frames. For all determined strain fields, the membrane DC pump intermittently pumps water flow into the PDMS structure from the left-end and then left from the right-end. Obviously, the left-ends of all cases present a high strain gradient that is because the water pressure changed due to the material properties difference between the PDMS and the plastic tubes. The other important phenomenon can be observed is that there is no significant change for both ε_{xx} and ε_{xy} strain map; however, the spatial distributions of ε_{xx} and ε_{xy} strain fields are various periodically along the horizontal direction. As for the ε_{yy} strain map, as shown in the last row of Fig. 7.6, the strain map present a high strain region which is almost perfect overlap with the 'artificial artery.' Besides, from left to right, the strain value in magnitude is increasing and then decreasing from the Nth frame to the N + 3th frame; the change of the strain value might also correspond to the dynamic behavior of water pumping but no strong evidence at this phase. However, based on the results, the artery below the skin might can be marked by using the strain map.

Figure 7.7 shows a series of strain map around the radial artery the high strain region can be easily identified, according to the results obtained from the artery simulator, the high strain region should be highly correlated with the radial artery. In comparison with the in-plane displacement map shown in Fig. 7.8, it fails to provide useful information such as for artery location.

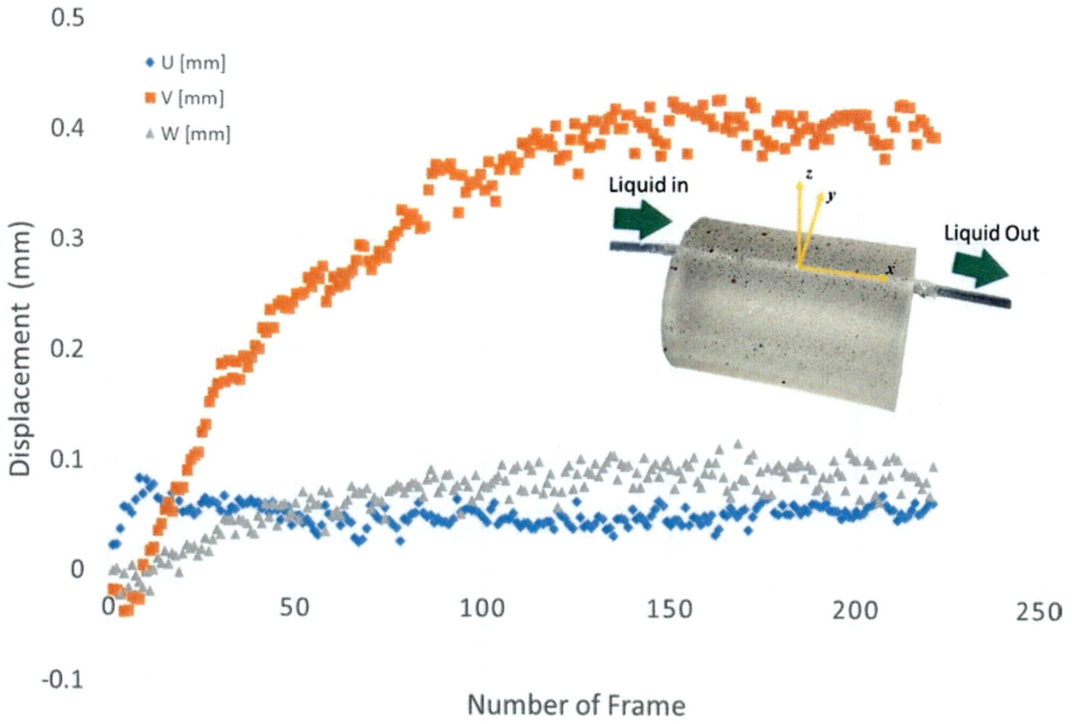

Fig. 7.5 Displacement of PDMS structure introduced by pumping water flow through

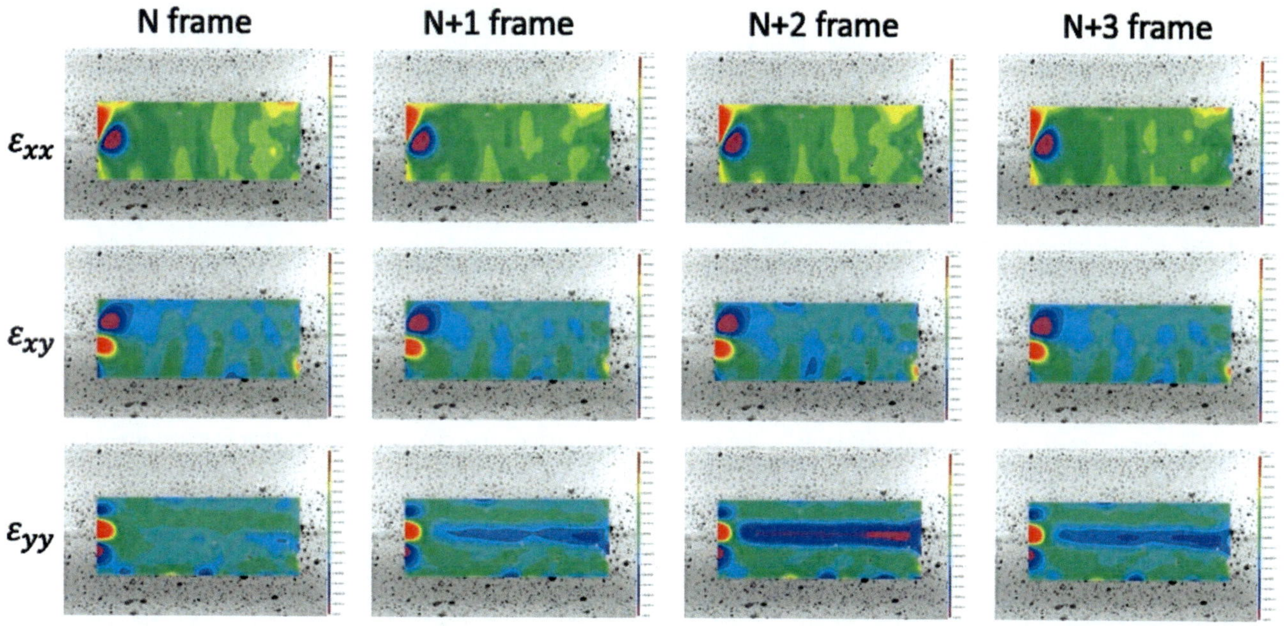

Fig. 7.6 Strain map above a tube of the PDMS structure

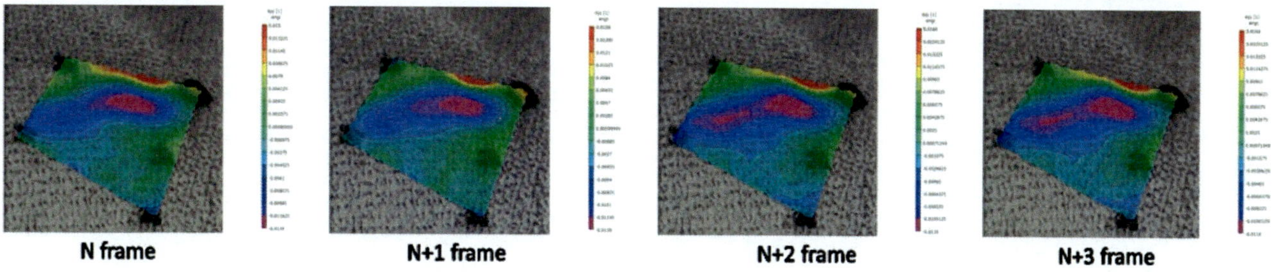

Fig. 7.7 A series strain map around the radial artery.

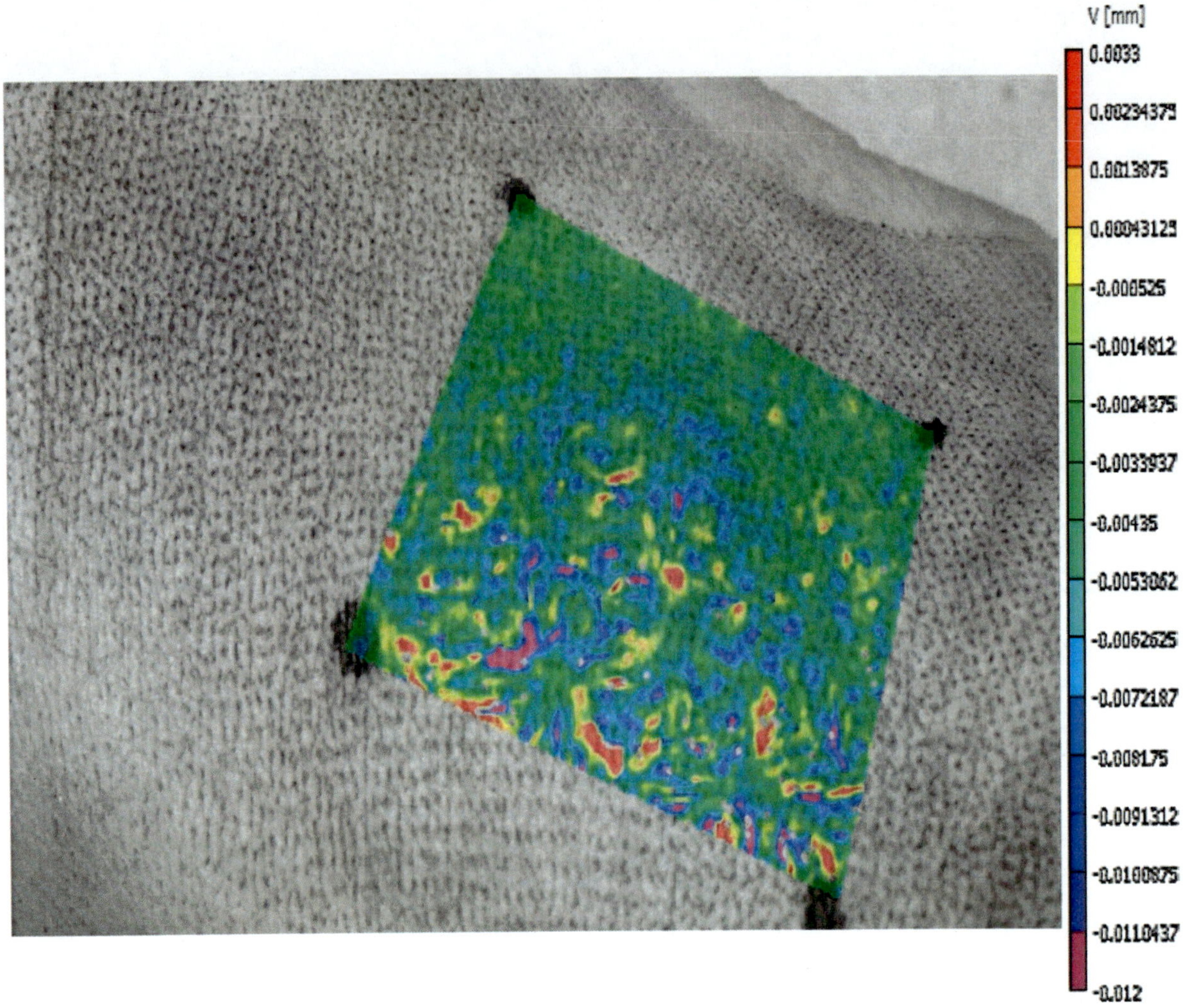

Fig. 7.8 The typical in-plane displacement around radial artery

7.4 Conclusions

In this study, a simplified artery vessel simulator is first developed. A PDMS hollow-cylinder with a through-depth hole below the outer surface is manufactured to simulate the artery locates inside an organ. The ECG signal is generated to drive the pump for drawing water out of a tank intermittently and forcing water to circulate through the model. Both displacement and strain maps around the simulation artery are determined by stereo-DIC. The results show that the high strain region is high correlates to the location of the simulation artery, especially for the displacement gradient normal to the artery. The displacement and strain field around the human radial artery present similar conclusions that support the result obtained by the developed simulator. Therefore, the artery beneath the skin can be identified by whole filed strain map which can be obtained by DIC method.

Acknowledgements This paper was supported in part by the Ministry of Science and Technology, Taiwan (Grand no. MOST-107-2221-E-492 -012 and MOST-108-2221-E-492-022-MY2).

References

1. Berfield, T.A., et al.: Fluorescent image correlation for nanoscale deformation measurements. Small. **2**(5), 631–635 (2006)
2. Xue, Y., et al.: High-accuracy and real-time 3D positioning, tracking system for medical imaging applications based on 3D digital image correlation. Opt. Lasers Eng. **88**, 82–90 (2017)
3. Solav, D., et al.: MultiDIC: An open-source toolbox for multi-view 3D digital image correlation. IEEE Access. **6**, 30520–30535 (2018)
4. Hokka, M., et al.: In-vivo deformation measurements of the human heart by 3D digital image correlation. J. Biomech. **8**(1), 2217–2220 (2015)
5. Hokka, M., et al.: DIC measurements of the human heart during cardiopulmonary bypass surgery Mechanics of Biological Systems and Materials, vol. 6, pp. 51–59. Springer, Cham (2016)
6. Soltani, A., et al.: An optical method for the in-vivo characterization of the biomechanical response of the right ventricle. Sci. Rep. **8**(1), 1–11 (2018)

Measuring Strain Distribution of Knee Cartilage Using Digital Volume Correlation

8

Maho Itakura, Yuelin Zhang, Kyohei Takeo, and Satoru Yoneyama

Abstract

The purpose of this study is to measure three-dimensional deformation of the porcine knee joint articular cartilage under compressive load by using digital volume correlation. The compression tester is construed for collecting the micro-CT images before and after the loading. The compressive load is applied to the test sample, and the compressive load is measured by a load cell. The micro-CT scanning is conducted, all of which are imported images. The three-dimensional displacement distribution and strain distribution of the articular cartilage are obtained using digital volume correlation. By examining the relationship between the compressive load and the strain distribution, the possibility of measuring the deformation characteristics of articular cartilage by using this method with minimum invasive is discussed.

Keywords

Digital volume correlation , Micro-CT , Knee cartilage , Strain distribution

8.1 Introduction

Articular cartilage is a connective tissue of joints, plays a role in load transfer, impact absorption and lubrication, and so is an important tissue to facilitate daily activities. Population of elderly people has been increasing all over the world due to progress in medical technology and the prevalence of advanced medical treatment. Thus, the number of cases of osteoarthritis (OA), a disease of articular cartilage, are rapidly increasing. OA develops when aging articular cartilage is worn down and damaged. To diagnose the disease at an early stage, a non-contact diagnostic method is required.

Diagnosis of cartilage damage is mainly performed on using computed tomography (CT). In this case, the damage of the cartilage is determined from the gap between the bones visually from the CT images. But, it is difficult to find it visually in the early stage when the damage area is small. Therefore, we focused on visualizing cartilage deformation. Normal and damaged cartilage deform differently under the same load. By observing the deformation by applying loads to the cartilage in various states and obtaining the relationship between the cartilage and the deformation, it is possible to predict the timing of the damage. As a method of visualizing the deformation, there is digital volume correlation [1] which can measure the displacement inside the object. Using DVC can calculate the displacement inside the human body [2], it may be possible to observe the local deformation of cartilage.

M. Itakura (✉) · K. Takeo · S. Yoneyama
Department of Mechanical Engineering, Aoyama Gakuin University, Sagamihara-shi, Kanagaea, Japan
e-mail: c5619112@aoyama.jp

Y. Zhang
Department of Engineering and Applied Sciences, Sophia University, Chiyoda-ku, Tokyo, Japan

J. Notbohm et al. (eds.), *Mechanics of Biological Systems and Materials & Micro-and Nanomechanics & Research Applications*, Conference Proceedings of the Society for Experimental Mechanics Series,
https://doi.org/10.1007/978-3-030-59765-8_8

In this research, the objective is to establish a less-invasive method to visualize the cartilage deformation under compressive load by using digital volume correlation.

8.2 Digital Volume Correlation

Digital volume correlation is a method of calculating the three-dimensional displacement inside the object by comparing the brightness value distribution within a subset of volume images before and after deformation. Illustration of digital volume correlation is shown as Fig. 8.1. The correlation coefficient C is used to calculate the brightness value distribution within the subset.

$$C\left(x,y,z,x^*,y^*,z^*\right)=\frac{\Sigma F\left(x,y,z\right)G\left(x^*,y^*,z^*\right)}{\sqrt{\Sigma F\left(x,y,z\right)^2\,\Sigma G\left(x^*,y^*,z^*\right)^2}} \tag{8.1}$$

Here, $F(x, y, z)$ is the brightness at the position (x, y, z) of the volume image before deformation, $G(x^*, y^*, z^*)$ is the brightness at the position (x^*, y^*, z^*) of the volume image after deformation. Σ means to calculate the sum of the brightness values within the subset. When the brightness value distribution within the same size subset of before and after deformation match, the correlation coefficient C reach the maximum. So, the displacement is obtained by calculating the position where the correlation coefficient C is the maximum. Also, when the transformation of the subset is not considered, (x^*, y^*, z^*) and (x, y, z) have the relationship of Eq. (8.2).

$$x^* = x + u_x$$

$$y^* = y + u_y$$

$$z^* = z + u_z \tag{8.2}$$

Here, u_x, u_y and u_z are displacement components in x, y and z directions, respectively. The strain distribution is calculated by the following Eq. (8.3) using the displacement distribution.

$$\varepsilon_x = \frac{\partial u_x}{\partial x} + \frac{1}{2}\left[\left(\frac{\partial u_x}{\partial x}\right)^2 + \left(\frac{\partial u_y}{\partial x}\right)^2 + \left(\frac{\partial u_z}{\partial x}\right)^2\right]$$

$$\varepsilon_y = \frac{\partial u_y}{\partial y} + \frac{1}{2}\left[\left(\frac{\partial u_x}{\partial y}\right)^2 + \left(\frac{\partial u_y}{\partial y}\right)^2 + \left(\frac{\partial u_z}{\partial y}\right)^2\right]$$

$$\varepsilon_z = \frac{\partial u_z}{\partial z} + \frac{1}{2}\left[\left(\frac{\partial u_x}{\partial z}\right)^2 + \left(\frac{\partial u_y}{\partial z}\right)^2 + \left(\frac{\partial u_z}{\partial z}\right)^2\right]$$

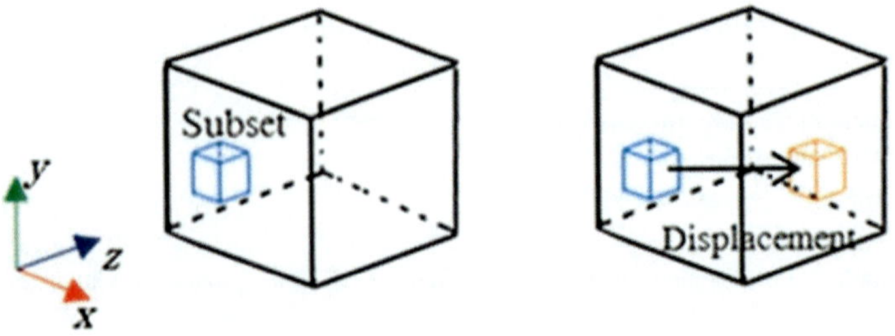

Fig. 8.1 Illustration of digital volume correlation

$$\gamma_{xy} = \frac{\partial u_y}{\partial x} + \frac{\partial u_x}{\partial y} + \frac{\partial u_x}{\partial x}\frac{\partial u_x}{\partial y} + \frac{\partial u_y}{\partial x}\frac{\partial u_y}{\partial y} + \frac{\partial u_z}{\partial x}\frac{\partial u_z}{\partial y}$$

$$\gamma_{yz} = \frac{\partial u_z}{\partial y} + \frac{\partial u_y}{\partial z} + \frac{\partial u_x}{\partial y}\frac{\partial u_x}{\partial z} + \frac{\partial u_y}{\partial z}\frac{\partial u_y}{\partial z} + \frac{\partial u_z}{\partial y}\frac{\partial u_z}{\partial z}$$

$$\gamma_{zx} = \frac{\partial u_y}{\partial z} + \frac{\partial u_x}{\partial x} + \frac{\partial u_x}{\partial z}\frac{\partial u_x}{\partial x} + \frac{\partial u_y}{\partial z}\frac{\partial u_y}{\partial x} + \frac{\partial u_z}{\partial z}\frac{\partial u_z}{\partial x} \tag{8.3}$$

8.3 Experiment

In this experiment, a compression test is performed using a porcine knee joint as a specimen. As shown in Fig. 8.2, the specimen size is about 90 mm including the femur and tibia of the left leg. The compression testing device is shown as Fig. 8.3. The image acquisition was performed using a micro-CT scanner (inspeXio SMX-100CT, SHIMAZU). The micro-CT scanning was conducted at 40 V and 200uA, all of which were imported as 16bit images with a size of 512 × 512 pixels. The resolution of the images was 0.089 × 0.089 × 0.089 mm^3/voxel. The captured images are shown in Fig.8.4a, b.

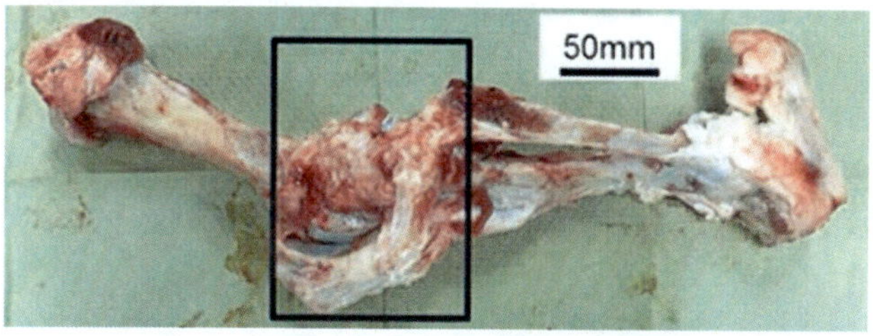

Fig. 8.2 The specimen of porcine knee joint

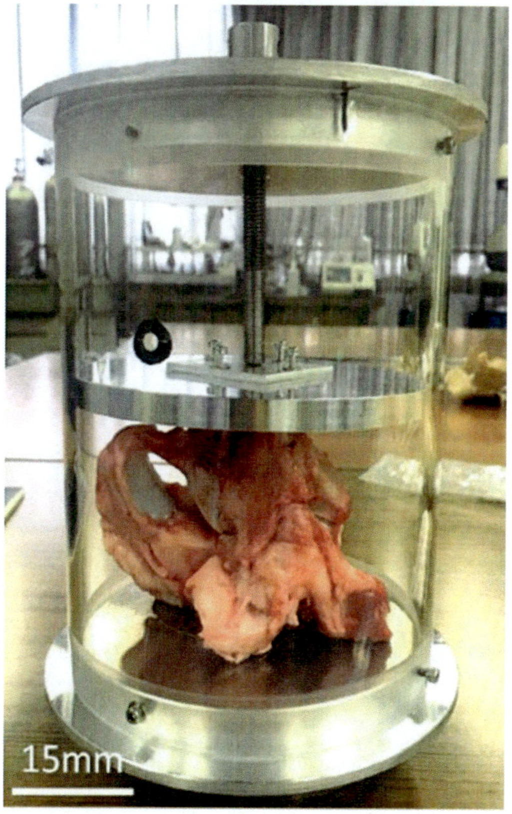

Fig. 8.3 The compression testing device

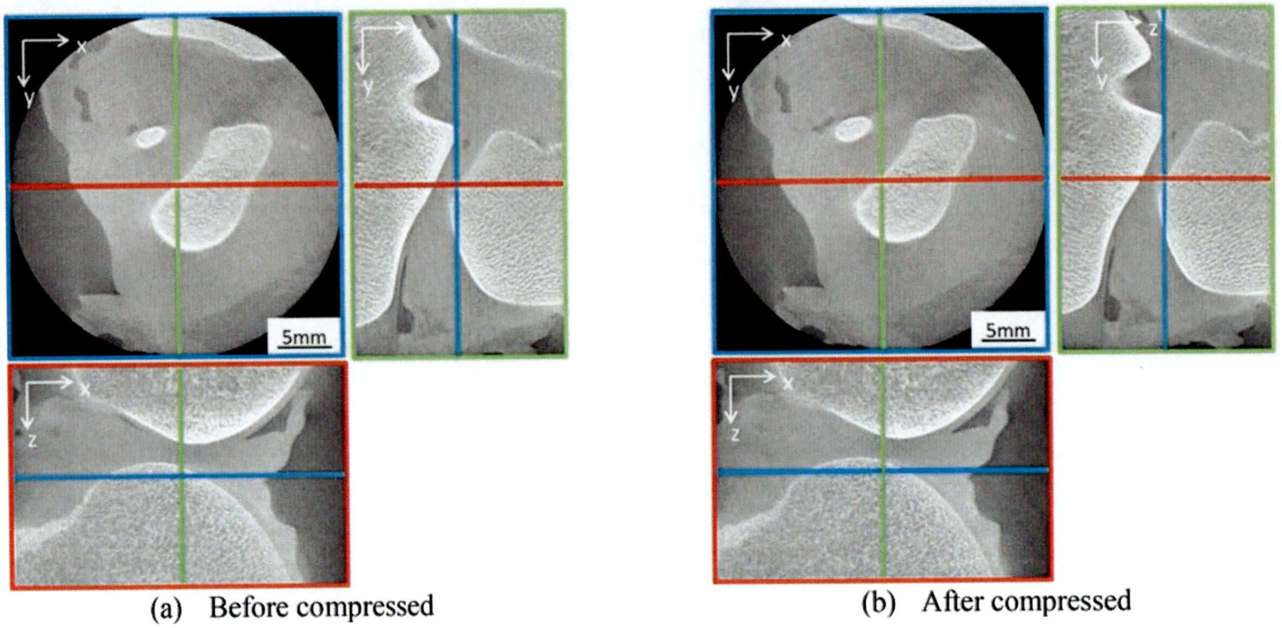

Fig. 8.4 Micro-CT image of the knee joint. (**a**) Before compressed. (**b**) After compressed

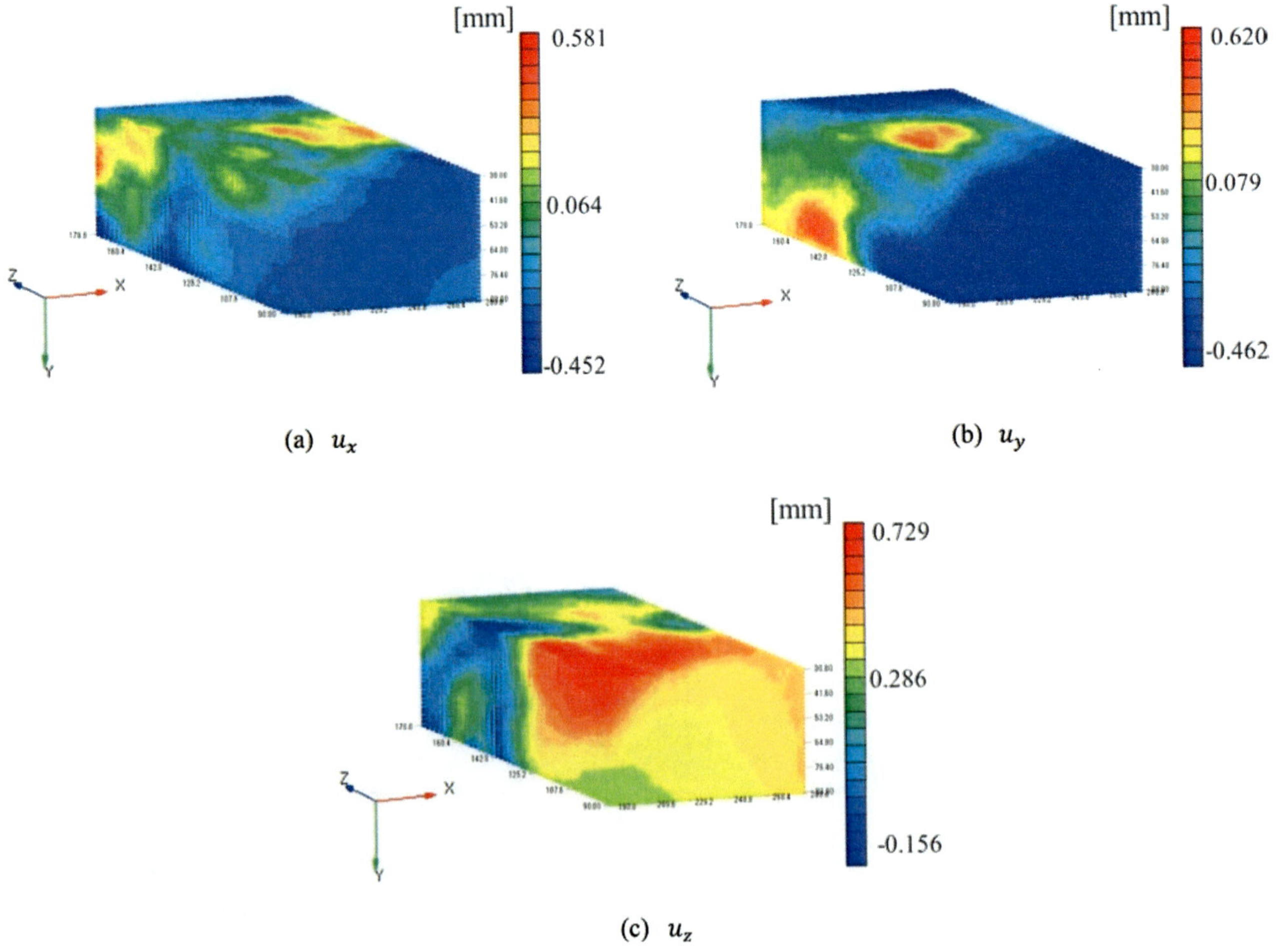

Fig. 8.5 Displacement distribution

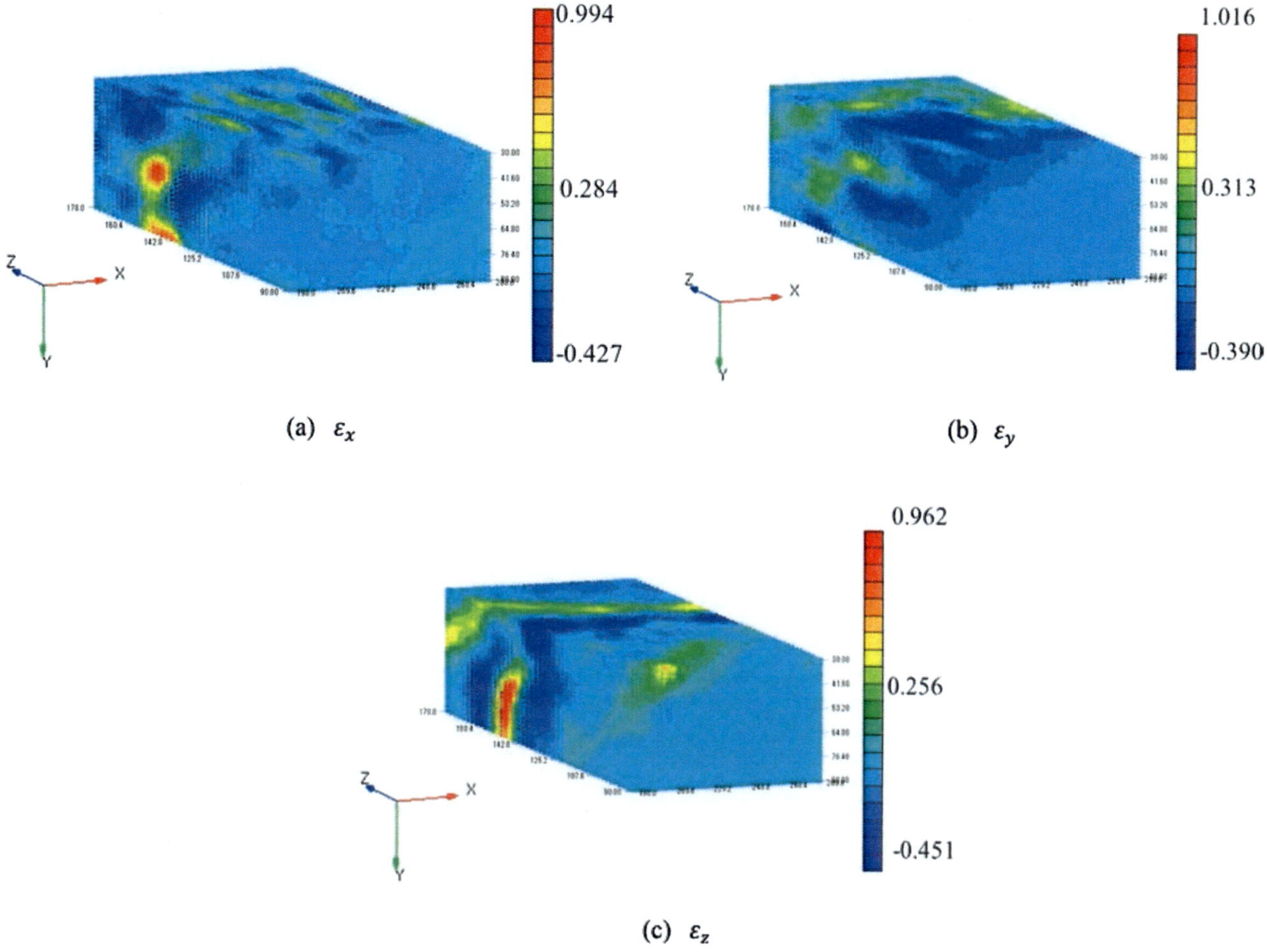

Fig. 8.6 Strain distribution

8.4 Distribution of Displacement and Strain

Digital volume correlation is performed on the micro-CT image to calculate the displacement and strain distribution. The analysis area is 8.9 × 5.3 × 8.0 pixels, including the cartilage where compression was confirmed. The subset size is 11 × 11 × 11 mm to 6.3 × 6.3 × 6.3 mm, and the search range is 1.86 × 1.86 × 1.86 mm.

The results of displacement distribution and strain distribution are shown as Figs. 8.5 and 8.6, respectively. In Fig. 8.5, the displacement was confirmed at the cartilage or meniscus. In each direction, but, both positive and negative displacements were obtained to the compression direction. The compression was performed uniaxial, so the cause is considered to be the gap between the bone and the tissue such as the meniscus. The results of ε_z in Fig. 8.6, the constant strain distribution occurs along the femur and tibia.

8.5 Conclusion

In this study, digital volume correlation was used to measure the three-dimensional deformation of porcine knee cartilage under compressive loading. Displacement and strain distributions were calculated from CT images before and after deformation. Deformation of the entire joint including cartilage could be visualized, but the deformation characteristics of the cartilage could not be evaluated. In the future, the deformation characteristics of the cartilage will be evaluated in consideration of the compression state.

References

1. Bay, B.K., Smith, T.S., Fyhrie, D.P., Saad, M.: Digital volume correlation: three-dimensional strain mapping using X-ray tomography. Exp. Mech. **39**, 217–226 (1999)
2. Kumazawa, S., Kuwazuru, O., Yoshikawa, N.: Three-dimensional displacement field identification within biological soft tissue. Seisankenkyu. **59**(2), 117–119 (2007)
3. Itakura, M., Zhang, Y., Takeo K,. Yoneyama, S.: "Measuring three-dimensional strain distribution of knee joint cartilage under compressive load", Proceedings of ATEM'19, 2019

9 Characterization of Shear Band Nucleation and Propagation in Bulk Metallic Glasses

Hui Wang and Ashraf Bastawros

Abstract

A novel experimental configuration was devised to monitor the deformation mechanisms in metallic glasses at the microscopic scale. The experiment is comprised of a wedge like cylindrical indenter while monitoring the evolution of the associated plastic zone, and crack initiation and propagation by a microscopic digital image correlation system. A new algorithm is developed to measure the localized shear strain with the shear band, which found to be in excess of several thousand percent. The deformation within the band and outside the band were resolved and showed very intricate cooperative interactions. The measurements and in-situ observations shed lights on the micromechanisms of deformation in BMG alloys.

Keywords

Bulk metallic glasses · Indentation · Digital image correlation · Shear bands

9.1 Introduction

Bulk metallic glasses (BMG) exhibit unique combination of high strength, large elastic strain limit (up to 2%) [1, 2], and high hardness [3]. Though, BMG fracture catastrophically by highly localized shear band in unconfined geometries, such as uniaxial tension or compression [1, 2, 4]. The BMG inhomogeneous plastic deformation of BMG tends to be highly localized into narrow shear bands. The macroscopic observations showed asymmetric deformation behavior between tensile and compressive loading [1] with moderate pressure dependent macroscopic yielding behavior [5–7]. Under geometric confinement, the BMG deformation exhibit increased ductility, accompanied with stable shear bands propagation [8]. Especially under inhomogeneous loading conditions during ribbon bending [9], and indentation field, either in 2D [6, 10, 11], or 3D [3, 5, 12, 13]. The microscopic view of deformation is viewed as a cooperative behavior of small clusters of randomly closed-packed atoms, so called shear transformation zones (STZs) [14], rather than the dislocation mechanisms in crystalline metals. The STZs are thought to create a localization of displacements in surrounding regions that triggers the evolution of highly localized shear bands during the deformation by creation of free volume [14]. Such phenomenological view has been further corroborated by molecular dynamic (MD) simulations [15, 16]. In continuum modeling, the free volume is considered to evolve with the applied stress [17]. The BMG plastic flow will occur when the free volume created by the applied stress exceeds the annihilation/diffusion rates. This microscopic view is further generalized [18, 19], to describe the initiation and propagation of shear bands. Transmission Electron Microscopy (TEM) analysis of the shear bands have indeed shown that the amount of voids has increased within a shear band, relative to surroundings [20].

H. Wang · A. Bastawros (✉)
Department of Aerospace Engineering, Iowa State University, Ames, IA, USA
e-mail: bastaw@iastate.edu

J. Notbohm et al. (eds.), *Mechanics of Biological Systems and Materials & Micro-and Nanomechanics & Research Applications*, Conference Proceedings of the Society for Experimental Mechanics Series,
https://doi.org/10.1007/978-3-030-59765-8_9

Realizing the differences in length scales between TEM observations, MD simulation and the continuum description of shear band nucleation and propagations, there is a lack in experimental observation of the microscopic deformation evolution for shear bands in BMG. In this work, we have analyzed the evolution of the shear band in BMG under constrained geometry. The wedge-like cylindrical indentation experiment, developed earlier [6, 11, 12] to allow in situ observation of the nucleation and propagation of shear bands, is utilized. The deformation field underneath the indenter is analyzed by DIC. A DIC numerical code has been developed based on subset method [21, 22] and the localized deformation filed has been explored by this technique. Summary of the material, experimental protocols and measurements of the shear strain within the shear band by DIC are given here. The details of the experimental results and the history of the deformation filed within the shear band and he surrounding matrix are derived from the measured fields.

9.2 Background

Experimentally, digital image correlation (DIC) should be a great tool to resolve such localized deformation [23, 24]. DIC technique has been developed for many years and it was first proposed in early 1980s to compute the surface strains and displacements [25]. It has been used either as a quantitative tool for average and homogeneous strains of macroscopically uniform deformation [26], or as a whole-field deformation mapping tool to measure highly localized deformation fields such as the ones surrounding a crack tip [27] and characterization of strain bifurcation in foams [22–24, 28, 29], doctor in details ahead of a particular shear band, detection of microscopic large strain by SEM topography image [30], plastic deformation patterns PLC bands [31], and shear band in solder joints [32].

As a deformation measurement technique, DIC provides estimation of the displacement field by correlating the features in a pair of digital images of a specimen surface before and after deformation by a mathematically well-defined function. DIC technique has different correlation algorithms, such as estimating discrete displacements data by local optimal match of intensity pattern in subsets (subset method) [33], or obtaining continuous displacement field function by full-field optimal match of the intensity pattern (continuous method) [34]. Different gray value interpolation schemes are employed; including the most common polynomial or B-spline for 2D surface [35] and tricubic interpolator based on 1D Hermitian cardinal basis functions for 3D volume [36]. The continuous method normally predicts accurate displacements and their gradients, when applied to a deformation field without localization [34]. However, it would smear out localizations, for highly inhomogeneous deformation fields, especially with discontinuities. Recently, for single shear band-like localization, an extended DIC approach has been introduced to partition the field into different parts and estimate the discontinuity [37]. In the highly localized heterogeneous deformation field, there would be multiple localizations and complex interaction with each other, such as the deformation field of bulk metallic glass under indentation [6, 10, 11]. The microscopic localization domain could be wider than a single line. Therefore, understanding the resolution and limitation for the highly localized deformation by DIC technique is critical to adjudge the width and strain level inside of the localization sites.

9.3 Experiment

A computer controlled Instron 8862 servoelectric loading frame is utilized for the cylindrical indentation test, with controlled crosshead displacement rate of 1 μm/s. A SiC wedge like cylindrical indenter with root radius, R of 2.4 mm is used in the experiment. The schematic of the developed loading fixture is shown in Fig. 9.1. The fixture ensures the alignment and symmetry of the indentation field. Additional details about the loading fixture, as well as systematic study of the self-similarity of the deformation field for other indenter radii can be found in Antonia el al. [6, 10, 11]. The experimental load-indentation depth curve is shown in Fig. 9.2a with the analysis stages marked on the curve. On the front surface of the sample, the evolution of the deformation zone underneath the indenter is in situ recorded during the course of loading by a traveling microscope with 5× objective lens and a progressive-scan camera (SPOT Insight CCD array of 2048x2048). A combination of through the lens and external oblique un-polarized white light was used to illuminate the specimen surface. Images are collected every 2 s during the indentation test. The field of view is 2 × 2 mm with a pixel resolution of 1 pixel/μm. This field of view is estimated to cover the entire process zone underneath the indenter up to the maximum applied load, based on estimates of the Johnson's (1970) cavity expansion model. A composite optical image showing the extent of the process zone at the end of the indentation test is shown in Fig. 9.2b, c.

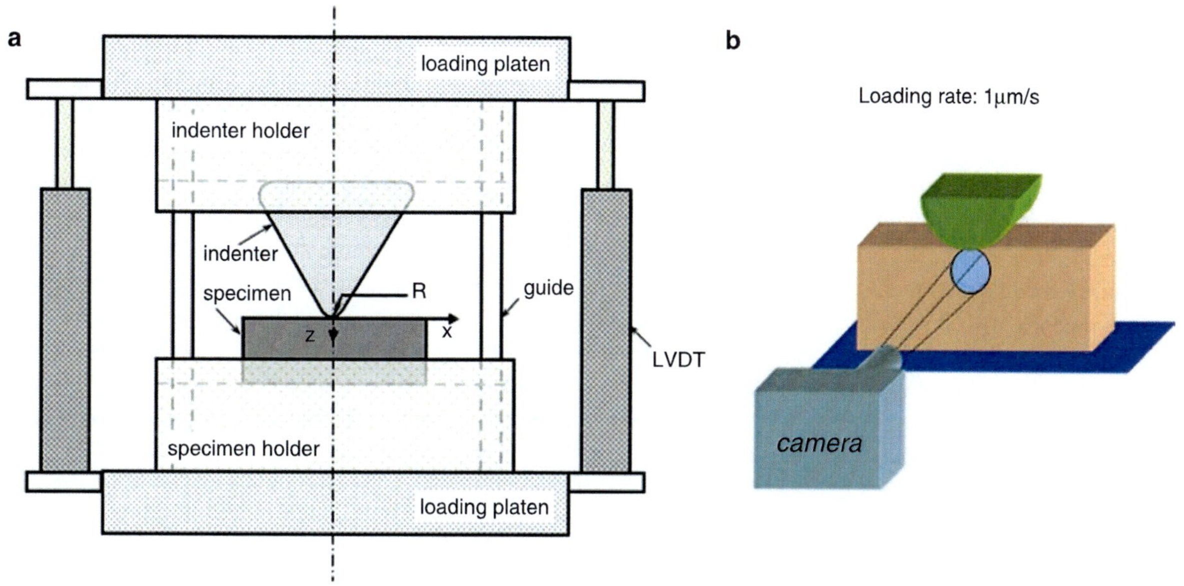

Fig. 9.1 Experimental Setup. (**a**) Schematic of the loading fixture for the cylindrical indentation testing. (**b**) Prospective of imaging setup

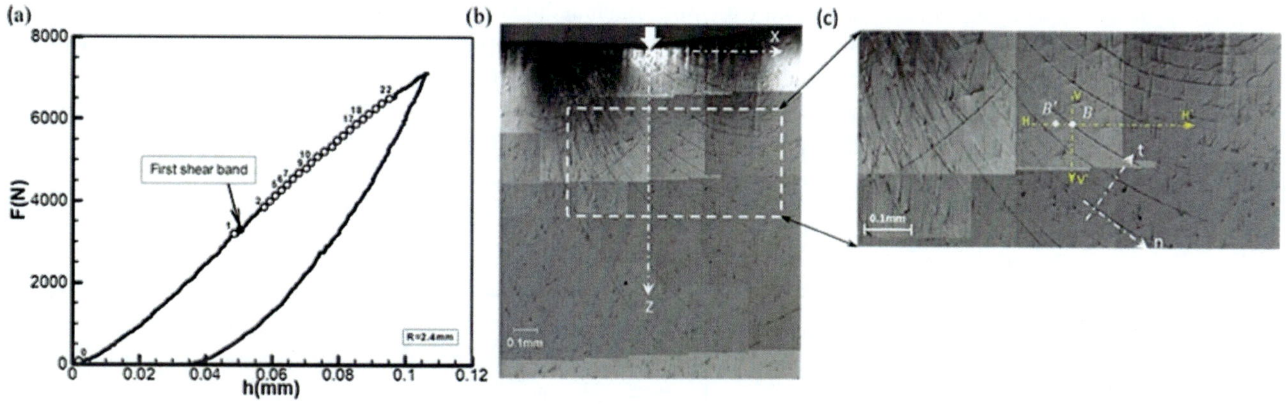

Fig. 9.2 (**a**) Experimental force-indentation depth response curve, showing the initiation of the shear bands and the analyzed stages of deformation. (**b**) A composite optical image of the indentation process zone after unloading. (**c**) Point of analysis along different shear bands

9.4 Strain Analysis

An in house program is used to analyze the deformation filed on the front surface underneath the indenter. Pairs of images of the current deformed configuration and the reference undeformed configuration are correlated together to find the nodal displacement over a uniform grid points of 5 pixel spacing. At each nodal point on the reference configuration, a correlation window of 31x31 pixels is correlated to the corresponding pair on the deformed configuration. The DIC analysis protocol has two-step, (i) a coarse search is used to approximately locate the nodal displacement to the nearest pixel [22, 33]. The search locates the correlation coefficients maxima within a search area (twice the correlation window size) on the deformed configuration. (ii) A fine search, utilizing the local intensity distribution, fitted to a two-dimensional third-order B-spline function is applied to minimize the error in the nodal displacement vector and their gradients. A direct strain evaluation from such displacement field would be meaningless as the gage length of calculation (approximately the window size) is several order of magnitude larger than the width of the shear band (of order of 10–20 nanometer [8]). The disparity between the shear band thickness and the required gage length to evaluate the displacement gradient precludes the ability of the DIC to accurately resolve the shear strain evolution within the band. The DIC technique overestimates the width of any deformation localization; for localization width that falls within the implemented gage length for each step of the analysis. For wider deformation localization, such as those associated with compaction bands in cellular solids and granular media, the effect of the analysis gage length can be corrected [24]. However, when the scale of the localization is much smaller than the analysis gage length, the strain estimates are significantly reduced.

Alternatively, if one accounts for the discontinuity in the displacement field across the shear band, an estimate of the displacement gradient can be found. Such estimate represents an average over the entire shear band thickness, t without considering the spatial shear strain distribution within the band. Though, t has to be known, which can be done independently after the indentation experiment. For displacement gradient analysis, selected points on the shear band, as depicted on Figs. 9.2c were chosen for the analysis at different stages of the whole loading cycle. The analysis starts by identifying the variation of the in-plane displacement components in two orthonormal directions, centered at the point of interest along the shear band, for all loading increments. The results of a selected points B in Fig. 9.2c on a shear band within the network, underneath the indenter will be presented here to show the evolution of the shear strain within the band, and in the neighboring matrix. Figure 9.3a–d shows the variation of the two in-plane displacement components (u, w) at point B for the incremental loading stage 5–6, along two orthonormal directions ($H - H'$ and $V - V'$). The $V - V'$ direction is chosen to be parallel to the indentation direction for ease of analysis. The displacement jumps ΔU_x, ΔW_x, ΔU_z and ΔW_z are estimated across the shear band, and marked on the figure. Estimates of the average displacement gradients across the shear band will be,

$$\begin{bmatrix} \dfrac{\partial u}{\partial x} & \dfrac{\partial u}{\partial z} \\ \dfrac{\partial w}{\partial x} & \dfrac{\partial w}{\partial z} \end{bmatrix} = \begin{bmatrix} \dfrac{\Delta U_x}{t/\sin\theta} & \dfrac{\Delta U_z}{t/\cos\theta} \\ \dfrac{\Delta W_x}{t/\sin\theta} & \dfrac{\Delta W_z}{t/\cos\theta} \end{bmatrix} \tag{9.1}$$

Where, θ is the orientation angle of the shear band with the horizontal direction. Once the component of the displacement gradient are defined on an average sense within the shear band, the in plane components of the Lagrangian strain components can be evaluated. Outside the shear band, such as at point B', the local displacement distribution is fitted to a bi-linear surface, to evaluate the spatial displacement gradients and the corresponding Lagrangian strain components within the shear band surrounding. To represent the severity of the deformation, the in-plane maximum total shear strain, $\gamma_{\max} = \sqrt{(\varepsilon_{xx} - \varepsilon_{zz})^2 + 4\varepsilon_{zx}^2}$ and the corresponding $\Delta\gamma_{\max}$ for loading increments will be utilized.

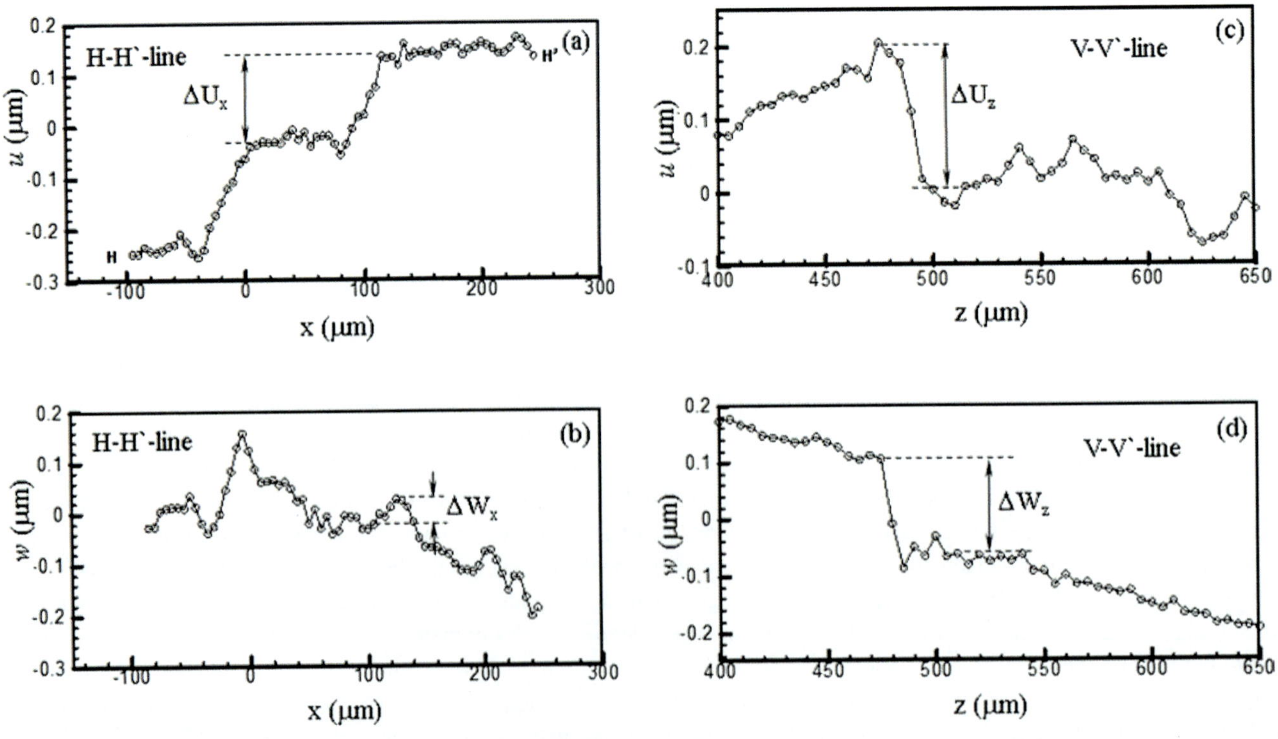

Fig. 9.3 Variation of the in-plane displacement components at point B of Fig. 9.2c, for the incremental loading stage 5–6, along two orthonormal directions H-H` and V-V` lines

An approximate width of the shear band can be calculated from the measured average strain components within the band. Assuming the local plastic volumetric strain to be conserved at the initiation of the shear band and neglecting the out of plane strain, then

$$\bar{\varepsilon}\Big|_{band}\left(\frac{t}{\lambda+t}\right)+\bar{\varepsilon}\Big|_{matrix}\left(\frac{\lambda}{\lambda+t}\right)=0 \tag{9.2}$$

Here, $\bar{\varepsilon}=\left(\varepsilon_{xx}+\varepsilon_{yy}\right)/3$, and λ is the spacing between bands. The mean strain values within and around the band are measured at the loading step, which corresponds to the first appearance of the shear band at point *B*. For the measured λ in this experiment, the shear band thickness *t* is estimated by Eq. (9.2) to be approximately 15 nm [8].

9.5 Results

The incremental change of the strain components for each of the monitored loading stages on Fig. 9.2a is calculated using the proposed algorithm for point, *B* marked on Fig. 9.2c. For each group of incremental strain components, $\Delta\gamma_{max}$ is evaluated employing Eq. (9.1). The average shear strain within the band, $\Delta\gamma_{band}$ and outside the band, $\Delta\gamma_{matrix}$ is summarized for each of the monitoring steps in Fig. 9.4a, and as a cumulative for the total activation cycle of the band in Fig. 9.4b. The evolution of $\Delta\gamma_{band}$ and $\Delta\gamma_{matrix}$ are plotted against a load derived shear strain measure, γ_{max} at the same spatial point. The γ_{max} represents the homogeneous maximum in-plane shear strain, derived from the Johnson's cavity expansion model [38]. Filled symbols represent activity within the shear band. It is apparent that there are close to three order of magnitude difference between the accumulated shear stain within the band and the surrounding matrix. During the activation of the shear band, the surrounding matrix does not show much activity, and vice versa, as indicated by the pairing of filled and unfilled symbols on Fig. 9.4. The shear band is active over several loading increment in Fig. 9.4a. Combining these consecutive increments together, according to their component, $\varepsilon_{ij}(n)=\varepsilon_{ij}(n-1)+\Delta\varepsilon_{ij}$ and then estimate $\Delta\gamma_{max}$, provides a measure of the cumulative shear strain within the band upon its incremental activation. The cumulative activation of $\Delta\gamma_{band}$ and $\Delta\gamma_{matrix}$ are shown in Fig. 9.4b. The matrix homogenous strain can now be estimated to be around 3.35–4%, before a shear band is nucleated. Such estimated level is consistent with homogeneous strain estimations of 4–6% based on cavity expansion model at the same load level. Upon the initiation of the shear bands, they propagate discontinuously with a significant increment of plastic shear

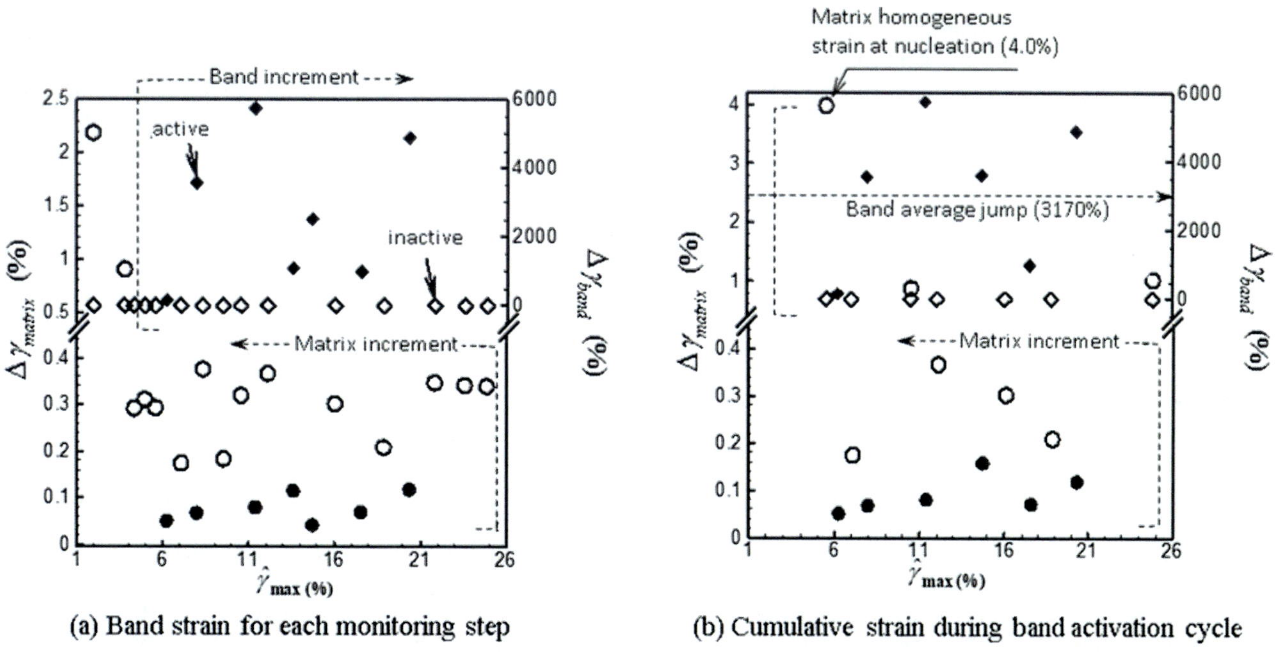

Fig. 9.4 Cumulative in-plane maximum shear strain increment, $\Delta\gamma_{max}$. Filled symbol-pairs for band and matrix represent band active stage. Unfiled symbol-pairs represent matrix active stage. (**a**) Band activation cycle shown to span multiple analysis increment. (**b**) Cumulative strain within the band when activated

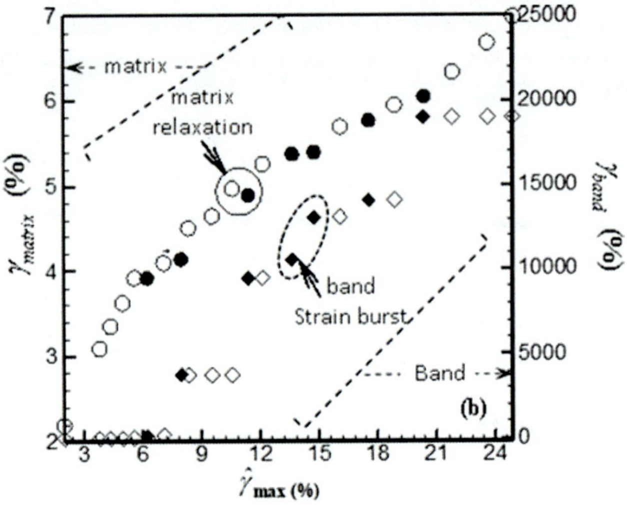

Fig. 9.5 Evolution of the total in-plane maximum shear strain increment, γ_{max}, averaged over the band width and the surrounding matrix

strain of about 3170–3550%. Such levels of strain can are three orders of magnitude higher than what direct estimates from DIC would indicate using continuum strain estimates.

Figure 9.5 shows the evolution of the total cumulative shear strain of γ_{band} and γ_{matrix} over the entire loading cycle. The total accumulated shear strain within the band reaches 18,000-19,000%. The surrounding matrix around the shear band has total averaged shear strain about 6–7%. On the figure, the filled symbols for the shear bands indicate activity. It is clearly seen that shear band deformation is highly discontinuous with incremental burst of the shear strain (filled symbols), followed by stationary stage (open symbols) until the macroscopic loading increases. During the stationary phase of the shear band, the matrix continues to deform with increasing the macroscopic loading. Moreover, the matrix deformation exhibited different strain rate during loading. However, the non-proportional strain path also exhibited the possibility of matrix strain relaxation during the shear band strain burst, as marked on Fig. 9.5.

9.6 Conclusion

A new analysis algorithm for the in-plane displacement field derived from DIC analysis is developed to show the evolution of the deformation field within the shear band in bulk metallic glasses. Many distinct features have been identified. The matrix was observed to reach a critical strain level of few percent before nucleation of a shear band. When the band is nucleated, it relax the strain within the surrounding matrix. The strain within the band reached several thousand percent. These measurements will provide a wealth of experimental result for direct calibration and comparison with continuum and multi-scale modeling framework.

Acknowledgements This work is supported by US National Science Foundation (NSF) with an award number CMS-0134111 and DMR-1807545.

References

1. Bruck, H.A., Christman, T., Rosakis, A.J., Johnson, W.L.: Quasi-static constitutive behavior of Zr41.25Ti13.75Ni10 Cu12.5Be22.5 bulk amorphous alloys. Scr. Metall. Mater. **30**, 429–434 (1994)
2. Wright, W.J., Schwarz, R.B., Nix, W.D.: Localized heating during serrated plastic flow in bulk metallic glasses. Mater. Sci. Eng. A. **319-321**, 229–232 (2001)
3. Donovan, P.E.: Plastic flow and fracture of Pd40Ni40P20 metallic glass under an indentor. J. Mater. Sci. **24**, 523–535 (1989a)
4. Pampillo, C.A.: Flow and fracture in amorphous alloys. J. Mater. Sci. **10**, 1194–1227 (1975)
5. Patnaik, M.N.M., Narasimhan, R., Ramamurty, U.: Spherical indentation response of metallic glasses. Acta Mater. **52**, 3335–3345 (2004)
6. Antoniou, A., Bastawros, A., Biner, B.: Experimental observations of deformation behavior of bulk metallic glasses during wedge-like cylindrical indentation. J. Mater. Res. **22**, 514–524 (2007)
7. Lund, A.C., Schuh, C.A.: Yield surface of a simulated metallic glass. Acta Mater. **51**, 5399–5411 (2003)

8. Kramer, M.J., Sordelet, D.J., Bastawros, A.F., Tan, X., Biner, S.B.: Absence of crystallization during cylindrical indentation of a Zr-based metallic glass. J. Non-Cryst. Solids. **351**, 2159–2165 (2005)
9. Conner, R.D., Johnson, W.L., Paton, N.E., Nix, W.D.: Shear bands and cracking of metallic glass plates in bending. J. Appl. Phys. **94**, 904–911 (2003)
10. Antoniou, A., Bastawros, A.F., Lo, C.C.H., Biner, S.B.: Deformation behavior of a zirconium based metallic glass during cylindrical indentation: In situ observations. Mater. Sci. Eng. A. **394**, 96–102 (2005)
11. Antoniou, A., Biner, S.B., Bastawros, A.F.: Experimental observation of cylindrical indentation of a metallic glass. Mater. Res. Soc. Symp. Proc. **903**, 0903-Z12-0 (2005)
12. Donovan, P.E.: Yield criterion for Pd40Ni40P20 metallic glass. Acta Metall. **37**, 445–456 (1989b)
13. Vaidyanathan, R., Dao, M., Ravichandran, G., Suresh, S.: Study of mechanical deformation in bulk metallic glass through instrumented indentation. Acta Mater. **49**, 3781–3789 (2001)
14. Argon, A.S.: Plastic deformation in metallic glasses. Acta Metall. **27**, 47–58 (1979)
15. Shi, Y., Falk, M.L.: Structural transformation and localization during simulated nanoindentation of a noncrystalline metal film. Appl. Phys. Lett. **86**(1), 011914-1-011914-3 (2005)
16. Shi, Y., Falk, M.L.: Does metallic glass have a backbone? The role of percolating short-range order in strength and failure. Scr. Mater. **54**, 381–386 (2006)
17. Spaepen, F.: Microscopic mechanisms for steady state inhomogeneous flow in metallic glasses. Acta Metall. **25**, 407–415 (1977)
18. Steif, P.S., Spaepen, F., Hutchinson, J.W.: Strain localization in amorphous metals. Acta Metall. **30**, 447–455 (1982)
19. Huang, R., Suo, Z., Prevost, J.H., Nix, W.D.: Inhomogeneous deformation in metallic glasses. J. Mech. Phys. Solids. **50**, 1011–1027 (2002)
20. Li, J., Spaepen, F., Hufnagel, T.C.: Nanometre-scale defects in shear bands in a metallic glass. Philos. Mag. A. **82**, 2623–2630 (2002)
21. Bruck, H., McNeill, S., Sutton, M., Peters, W.I.: Digital image correlation using Newton-Raphson method of partial differential correction. Exp. Mech. **29**, 261–267 (1989)
22. Wang, Y., Cuitiño, A.M.: Full-field measurements of heterogeneous deformation patterns on polymeric foams using digital image correlation. Int. J. Solids Struct. **39**, 3777–3796 (2002)
23. Bastawros, A., McManuis, R.: Case study: use of digital image analysis software to measure non-uniform deformation in cellular aluminum alloys. Exp. Tech. **22**(2), 35–37 (1998)
24. Wang, H., Lai, W., Antoniou, A., Bastawros, A.: Application of digital image correlation for multiscale biomechanics. In: Handbook of imaging in biological mechanics. Corey Neu and Guy Genin EDs, pp. 141–150. CRC Press, Boca Raton, FL (2014)
25. Peters, W.H., Ranson, W.F.: Digital imaging techniques in experimental stress analysis. Opt. Eng. **21**, 427–431 (1982)
26. Chu, T.C., Ranson, W.F., Sutton, M.A., Peters, W.H.: Applications of digital image correlation techniques in experimental mechanics. Exp. Mech. **25**, 232–244 (1985)
27. Antoniou, A., Onck, P., Bastawros, A.F.: Experimental analysis of compressive notch strengthening in closed-cell aluminum alloy foam. Acta Mater. **52**, 2377–2386 (2004)
28. Bastawros, A., Bart-Smith, H., Evans, A.: Experimental analysis of deformation mechanisms in a closed-cell aluminum alloy foam. J. Mech. Phys. Solids. **48**, 301–322 (2000)
29. Bastawros, A., Evans, A.G.: Deformation heterogeneity in cellular al alloys. Adv. Eng. Mater. **2**(4), 210–214 (2000)
30. Kang J., Schmidt T., Jain M., Wilkinson D.: Microscopic material characterization using SEM topography images correlation. SEM Annual Conference and Exposition on Experimental and Applied Mechanics 2006, Jun 4–7 2006, pp. 1389–1395
31. Tong, W.: Detection of plastic deformation patterns in a binary aluminum alloy. Exp. Mech. **37**, 452–459 (1997)
32. Antoniou, A., Bastawros, A.: Deformation characteristics of tin-based solder joints. J. Mat. Res. **18**(10), 2304–2309 (2003)
33. Sutton, M.A., Wolters, W.J., Peters, W.H., Ranson, W.F., McNeill, S.R.: Determination of displacements using an improved digital correlation method. Image Vis. Comput. **1**, 133–139 (1983)
34. Cheng, P., Sutton, M.A., Schreier, H.W., McNeill, S.R.: Full-field speckle pattern image correlation with B-spline deformation function. Exp. Mech. **42**, 344–352 (2002)
35. Schreier, H.W., Braasch, J.R., Sutton, M.A.: Systematic errors in digital image correlation caused by intensity interpolation. Opt. Eng. **39**, 2915–2921 (2000)
36. Bay, B., Smith, T., Fyhrie, D., Saad, M.: Digital volume correlation: three-dimensional strain mapping using X-ray tomography. Exp. Mech. **39**, 217–226 (1999)
37. Rethore, J., Hild, F., Roux, S.: Shear-band capturing using a multiscale extended digital image correlation technique. Comput. Methods Appl. Mech. Eng. **196**, 5016–5030 (2007)
38. Johnson, K.L.: Correlation of indentation experiments. J. Mech. Phys. Solids. **18**, 115–126 (1970)

Design of Purely Mechanical Device to Aid in the Suppression of Parkinsonian Hand Tremors

10

Kristen LeBar, Nadine Prentice, and Timothy Doughty

Abstract

Parkinson's disease (PD) is a neurodegenerative disorder that is caused by a decrease and lack of dopamine production in the brain. Patients experience inhibited mobility—particularly in their hands, which is where the tremors occur most often. There is a need for a tremor suppressor to improve patients' quality of life. The research conducted to develop a purely mechanical device to aid in the suppression of the hand tremors, as well as the details regarding the theory, prototyping, and testing will be presented. Other solutions such as electronic devices, brain stimulation and weights are invasive, cumbersome, and can be expensive. The mechanical device offers a noninvasive, affordable solution to the tremors. While other devices are designed to react to the tremors, the current device in development suppresses the tremors, thereby allowing the patient to have full control of movement. It is designed to be adjustable, giving it the ability to suppress over a range of patient tremors, which vary from 4 Hz to 7 Hz.

Keywords

Mechanical device, Resonate frequency, Tremor suppression

10.1 Introduction

Parkinson's disease is caused by a lack of dopamine production in the brain. The deficit of this neurotransmitter is caused by nerve cell damage, and it results in a severe effect on patients' motor skills. This leads to slower and stiffer movements, tremors predominantly affecting the hands, a loss of automatic movements, and changes in speech [1]. As the disease progresses, patients may have problems such as depression, trouble sleeping, and difficulties with chewing or swallowing [2]. Each year, approximately 60,000 Americans are diagnosed with Parkinson's disease. By 2020, one million Americans will be living with PD [3].

The tremors hinder the patients' ability to complete daily tasks, such as eating, drinking, dressing themselves, writing and typing. Parkinson's patients have very little independence because of these limitations. This may cause a decrease in self-confidence, and patients are often self-conscious of their hand tremors. However, with a tremor suppressing device, patients will be given the ability to complete these daily tasks on their own. With no known cure for the disease, PD patients often resort to surgery or medication. This research project focused on designing a purely mechanical device that suppresses the tremors in hopes of making daily activities less challenging for those with Parkinson's disease.

The tremor suppression device in development is a wearable device that the patient can have on for all activities and daily tasks, thus making their daily routines less challenging. Because it is a purely mechanical device, there is no need for external

K. LeBar · N. Prentice · T. Doughty (✉)
Shiley School of Engineering, University of Portland, Portland, OR, USA
e-mail: doughty@up.edu

J. Notbohm et al. (eds.), *Mechanics of Biological Systems and Materials & Micro-and Nanomechanics & Research Applications*, Conference Proceedings of the Society for Experimental Mechanics Series,
https://doi.org/10.1007/978-3-030-59765-8_10

power. Furthermore, water does not affect the device, therefore it can be worn when washing dishes, washing hands, or bathing. The mechanical components of the device are also made with low cost materials, making it a feasible solution for any patient.

10.2 Background

Previous tremor devices designed at other institutions have stabilized, measured, and sometimes suppressed the tremors for Parkinson research. Some devices use sensors embedded in patients' clothing in an attempt to read the tremors and stabilize the inputted frequencies [3]. However, some sensors are even embedded in patients' brains. The latter option is extremely invasive and sometimes not feasible for older patients that cannot undergo surgery. The wearable mechanical device described in this paper will suppress the tremors and will not require recalibration. Patients' tremors are not violent enough to damage the device, so the mechanical device could be used for a long period of time.

The team is working to develop a purely mechanical device with capabilities to be adjusted to suppress Parkinsonian hand tremors at a range of frequencies. Previous devices from the research team have also suppressed the tremors ranging from 4 Hz to 7 Hz. The utensil developed in 2012 [4] suppressed the tremors at 7 Hz. For the purposes of the wearable device, the team is designing it to suppress the tremors between 4 and 7 Hz. Tremors of the resting hand lie primarily at 5 Hz, although the range is still 4–7 Hz. A tremor suppression device is more applicable for patients with Parkinson's disease. A stabilizing device does not stop the tremors, it counteracts them meaning the tremors will still be there, but the patient cannot feel them because the device is working at opposite frequencies. These devices have revolutionized both how patients are treated and what style of utensils and devices they use.

10.3 Methods

The experimental setup modeled an actual human hand system. The test hand mass was 400 g, approximately the same mass as an actual human hand – 406 g – and spring stiffness of the system was such that the system resonated at 5 Hz, the same as the natural frequency of a human hand.

On the control side of the system, a dynamic shaker was used to initiate the motion of the testing hand. LabVIEW ran a program to run a sweep test, stimulating the system from 2 Hz to 10 Hz, with 31 steps. This range was chosen because the desired resonate frequency lies in this range, and the behavior of the system on either side of the resonate frequency could be seen on the response diagram. The sweep test gave more accurate results because LabVIEW could settle at the desired frequency, giving a clearer frequency response.

From LabVIEW, the code was run to the data acquisition assistant (DAQ), which is the hardware used to receive and give data to LabVIEW. From the DAQ, the signal was sent to the amplifier which controlled the dynamic shaker. The shaker was a piston arm attached to a drum that oscillated at a steady frequency. The piston had a link connection at the center, offering a greater range of motion. Running through each frequency, the shaker, which was directly connected to the testing hand via the piston arm, stimulated the human hand system. When a device was tested, it was placed directly on the back of the test apparatus hand, simulating the placement of it on a patient's hand. The testing setup is given in Fig. 10.1 below.

On the measurement side, the arm was only attached to what the team called "stimulus side." On the stimulus side was a 2D accelerometer which measured the acceleration of the oscillation. The team decided a 2D accelerometer was sufficient for the stimulus side because the response side of the system was giving the vital data. The tremors in an actual human hand are three-dimensional, so a 3D accelerometer was placed on the response side of the system to read that data. Both accelerometer readings were sent through a 30 Hz low-pass filter to refine the data. From the filter the data was sent back to the DAQ and plotted on a live frequency response diagram in LabVIEW. Where there was a peak on the frequency response diagram was the frequency at which the system was resonating. See Fig. 10.2 for an illustration of the experimental procedure.

Characteristics of the materials used for device prototyping were measured and quantified to determine the frequency at which it would suppress the tremors. Details regarding materials used and measurement techniques are proprietary and cannot be stated. Before a prototype was tested, the system was run without a device to ensure the testing apparatus was resonating at 5 Hz (i.e. modeling an accurate human hand system). For testing, the device prototype was placed on the test hand.

Fig. 10.1 Testing setup

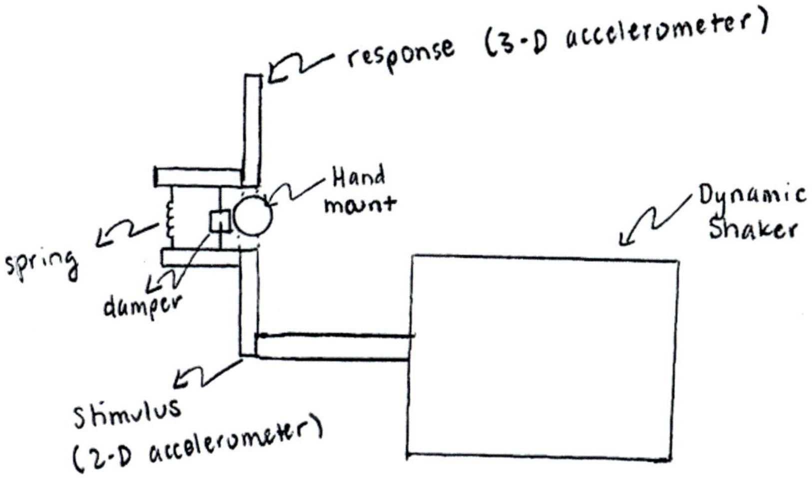

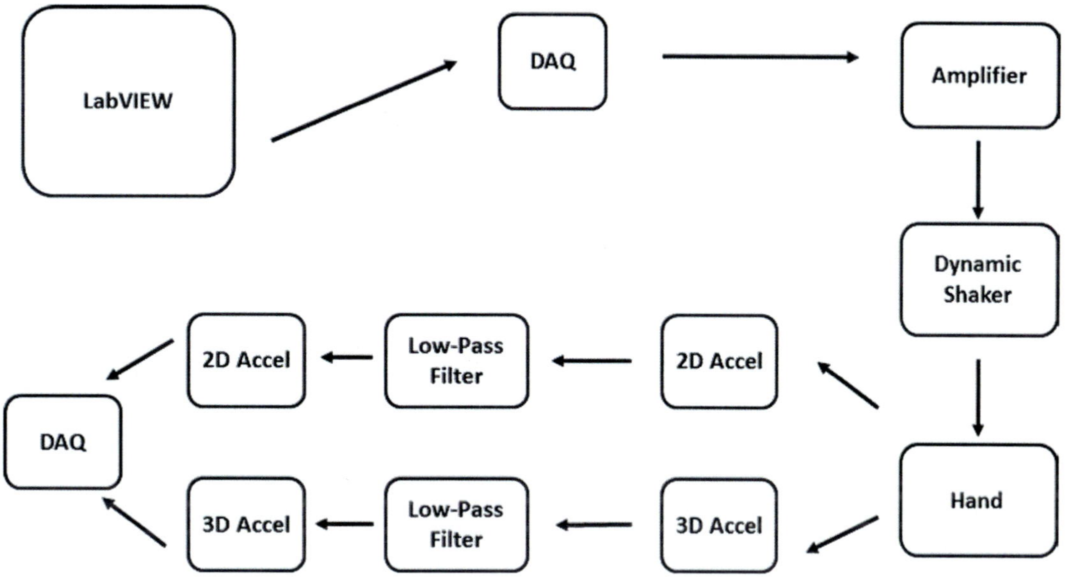

Fig. 10.2 Diagram of testing procedure-data flow

Due to the nature of the material used, the frequency at which it suppressed tremors decreased with successive runs. Therefore, runs for an individual prototype would continue until the frequency at which an antinode occurred was consistent across runs.

To assess the viability of an adjustability component, when the antinode became consistent, the prototype components were adjusted to suppress at a different frequency. It was most feasible to adjust the components such that the frequency at which the device suppressed increased. Because the range of tremors is 4–7 Hz, the device was designed to suppress at the lower end of that range so that it could be adjusted and suppress at frequencies above 4 Hz.

10.4 Results

Tremor suppression device iterations began to suppress tremors at 4 Hz. As stated before, successive runs (without any adjustments) decreased the frequency at which an antinode was created. Devices were developed and tested until they suppressed tremors at 4 Hz consistently. Adjusting the device from this point resulted in suppressing over the desired range (4–7 Hz) of frequencies. See Fig. 10.2 for the series of frequencies one iteration suppressed at.

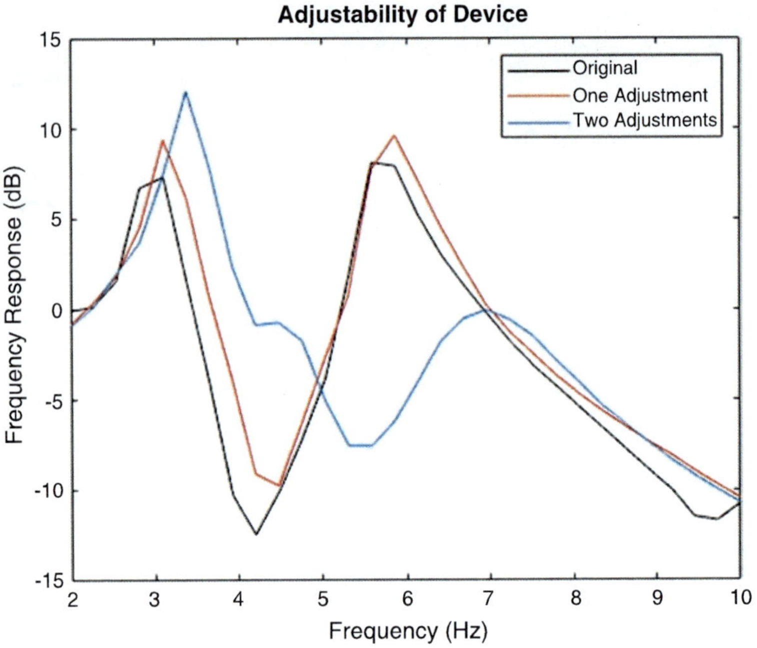

Fig. 10.3 One device iteration suppressing at three different frequencies with adjustments

Table 10.1 Summary of tremor suppression devices for Parkinson's disease

Author(s)	Year	Type of device	Summary
Doughty [4]	2012	Utensil tremor suppressor	Effective-antinode at 7 Hz
Bonato [5]	2005	Embedded sensors	Monitor tremors and measure frequencies
Loureiro, Belda-Lois, Lima, Pons, Sanchez-Lacuesta, Harwin [2]	2005	Tremor suppressor—controllable double viscous beam actuator	Damping action in response to tremor frequency
J.A. Gallego, E. Rocon, J. Ibánez, J.L. Dideriksen, A.D. Koutsou, R. Paradiso, M.B. Popovic, J. Belda-Lois, F. Gianfelici, D. Farina, D.B. Popovic, M. Manto, T. D'Alessio, and J.L. Pons [1]	2011	Wearable robot- tremor assessment and suppression	First iteration showed promise in assessing and suppressing tremors
M.C. Rodriquez-Oroz, J.A. Obeso, A.E. Lang, J.-L. Houeto, P. Pollak, S. Rehncrona, J. Kulisevsky, A. Albanese, J. Volkmann, M.I. Hariz, N.P. Quinn, J.D. Speelman, J. Guridi I. Zamarbide, A. Gironell, J. Molet, B, Pascual-Sedano,B. Pidoux, A.M. Bonnet, Y. Agid, J. Xie, A.-L. Benabid, A.M. Lozano, J. Saint-Cyr, L. Romito, M.F. Contarino, M. Sceratti, V. Fraix, N. Van Blercom [3]	2005	Brain stimulation	6–12 months of deep brain stimulation improves severe mobility complications
Present work K. LeBar, N. Prentice, T. Doughty PhD	2019	Mechanical wearable device-tremor suppressor with adjustability component	Created series of antinodes at 4.25, 4.5 and 5.25 Hz

10.5 Conclusion

The results of this research were fairly satisfactory. The team was able to suppress tremors over a range of tremors from 4 Hz to 7 Hz. In Fig. 10.3, the frequency responses are shown shifted based on the team's adjustments. The team used a mass-spring-damper system to model a human hand which, because the system was resonating at 5 Hz ± 0.5 Hz, accurately represented an average Parkinsonian tremor. The apparatus has remained the same through multiple iterations of research meaning the results demonstrate the device's effectiveness when compared to past and future iterations.

To further the research, reducing the overall device size and simplifying the adjustability component will be necessary. First, the device measures 2 in^3 which is too large to sit atop a hand and therefore the components need to be reworded to save space within the device. The current housing also does not have an effective means of attaching to the back of the patient's hand. Determining the attachment method will alter the current device size. Second, the adjustability competent involves removing and altering the internal elements of the device which is problematic. The patient will not be able to adjust the parts on their own and therefore the goal of the project – to create allow for a self-sufficient individual – is not being met. The adjustability competent will need to be researched and prototyped further and will have an effect on the device size mentioned previously.

Despite the need for further research, the current suppression device has the potential to improve a patient's quality of life by allowing them to perform individualized tasks and establish better hand mobility. This research will continue to benefit patients with Parkinsonian disease and bring them one step closer to living a more independent life.

Acknowledgements Special thanks to the Engineering Project Development Fund and Shiley School of Engineering Endowment for funding and making this project possible. Thank you to the team's advisor Dr. Timothy Doughty for his guidance and support for the project and passion for the research topic. Additionally, thank you to Jared Rees, Jacob Amos, and Christina Chrestos for their assistance in the machine shop and to Allen Hansen for his oversight in the Pilot MakerSpace.

References

1. Gallego, J.A., Rocon, E., Ibanez, J., Dideriksen, J.L., Koutsou, A.D., Paradiso, R., Popovic, M.B., Belda-Lois, J.M., Gianfelici, F., Farina, D., Popovic, D.B., Manto, M., D'Alessio, T., Pons, J.L.: A soft wearable robot for tremor assessment and suppression. http://digital.csic.es/bitstream/10261/43664/3/soft_wearable_robot_gallego_et_al.pdf (2011). Accessed 27 Sept 2018
2. Loureiro, R., Belda-Lois, J., Lima, E., Pons, J., Sanchez-Lacuesta, J., Harwin, W.: Upper limb tremor suppression in ADL via an orthosis incorporating a controllable double viscous beam actuator. In: Proceeding of the 2005 IEEE 9th International Conference on Rehabilitation Robotics, Chicago, IL, USA, 28 June–1 July 2005
3. Rodriquez-Oroz, M.C., Obeso, J.A., Lang, A.E., Houeto, J.-L., Pollak, P., Rehncrona, S., Kulisevsky, J., Albanese, A., Volkmann, J., Hariz, M.I., Quinn, N.P., Speelman, J.D., Guridi, J., Zamarbide, I., Gironell, A., Molet, J., Pascual-Sedano, B., Pidoux, B., Bonnet, A.M., Agid, Y., Xie, J., Benabid, A.-L., Lozano, A.M., Saint-Cyr, J., Romito, L., Contarino, M.F., Sceratti, M., Fraix, V., Van Blercom, N.: Bilateral deep brain stimulation in Parkinson's disease: a multicenter study with 4 years follow-up. Brain. **128**(Pt 10), 2240–2249 (2005) https://academic.oup.com/brain/article/128/10/2240/274634
4. Doughty, T., Heintz, J., Ishii, M.: Reducing Parkinsonian Hand Tremor with a Novel Dynamic Eating Utensil. In: Proceeding of the 2013 ASME International Engineering in Congress and Exposition, San Diego, CA
5. Bonato, P.: Advances in wearable technology and applications in physical medicine and rehabilitation. J. NeuroEngineering Rehabil. **2**, 2 (2005). https://jneuroengrehab.biomedcentral.com/articles/10.1186/1743-0003-2-2. Accessed 27 Sept 2018

Check for updates

Biomechanical Testing of Human Red Blood Cells Under Controlled Oxygen Tension

11

E. Du and Yuhao Qiang

Abstract

Red blood cells are subjected to dynamic loads of shear and oxidative stresses when they traverse through the blood circulation system. Cell membranes are not only subjected to mechanical forces but also oxidative damage as a result of hypoxia. We develop an experimental strategy that can subject biological cells to well-controlled shear and gaseous microenvironment in a microfluidic device. Significant changes in cell deformability and relaxation characteristics due to variation in oxygen tension are observed and quantified. Our results demonstrate that hypoxia can lead to significant loss of deformability in red blood cells. The relationship between the membrane fatigue degradation and the repeated stressing conditions of blood circulation can be further investigated using the developed platform.

Keywords

Cell biomechanics, Red blood cells, OXYGEN tension, Hypoxia, Microfluidics

11.1 Introduction

A key biological function of red blood cells (RBCs) is to carry oxygen to the body tissues and carbon dioxide away from the tissues. They are subjected to severe mechanical filtering process in small capillaries and sinusoidal walls of reticuloendothelial organs during their journey in blood circulation. The membrane of RBC is the only structural component, accounting for all of its diverse antigenic, transport, and mechanical characteristics [1]. RBCs have remarkable deformability. The repeated filtering process and oxidative stress gradually impair the membrane integrity and their ability in oxygen delivery as well as induce cell aging, leading to rigid RBCs for degradation by the mononuclear phagocyte system [2]. Biomechanics of RBCs and the sub-cellular components have been studied extensively for decades, using classical experimental strategies such as atomic force microscopy, micropipette aspiration, optical tweezers etc. [3–6]. Majority of these techniques are not well adapted in providing a well-controlled gaseous microenvironment of cells while perform biomechanics characterization. Therefore, to explore the interactions of cell biomechanics and oxidative stress, e.g. induced by hypoxia, an experimental methodology is needed.

E. Du (✉) · Y. Qiang
Department of Ocean and Mechanical Engineering, College of Engineering and Computer Science, Florida Atlantic University, Boca Raton, FL, USA
e-mail: edu@fau.edu

J. Notbohm et al. (eds.), *Mechanics of Biological Systems and Materials & Micro-and Nanomechanics & Research Applications*, Conference Proceedings of the Society for Experimental Mechanics Series,
https://doi.org/10.1007/978-3-030-59765-8_11

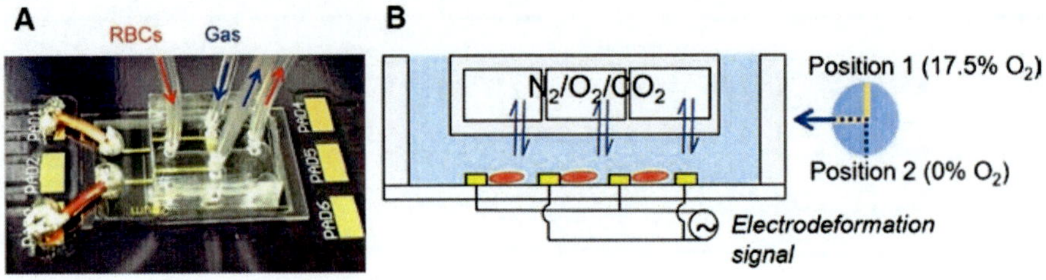

Fig. 11.1 (**a**) Microfluidic device for electrodeformation-based biomechanical testing of cells under controlled oxygen tension conditions. (**b**) Schematic of the cross-section of the microfluidic device for performing electrodeformation excitation of cells under well-controlled gaseous microenvironment with 17.5% O_2 (high oxygen tension) and 0% O_2 (low oxygen tension)

11.2 Background

Microfluidics technology has significantly advanced the study of blood flow and vascular physiology and pathology. These microfluidics-based platforms, typically with a geometrically constrained microchannel or channel network [7], enable accurate replication of vascular systems, in terms of geometry, dimension, and mechanics. Microfluidics is also a versatile platform, allowing integration of multiple functional units to the geometric constraints. A double-layer microfluidic channel was developed for precise control of the gaseous microenvironment of cells for the study of morphological and rheological changes in sickle blood cells [8, 9]. Benefited from the advanced microfabrication technique, thin film microelectrodes can be patterned to the substrate of microfluidic devices, for quantitative biomechanical characterization of RBCs via electrodeformation technique [10–12].

This paper presents an experimental strategy that integrates the electro-deformation technique and the hypoxia environment into a microfluidic platform for the biomechanical testing of cells in suspension. The microfluidic device consists of a pair of integrated gold microelectrodes for electrodeformation testing and a double layer structure made of Polydimethylsiloxane (PDMS), as shown in Fig. 11.1. The double layer is comprised of a gas channel atop a cell channel, and an in-between gas-permeable film of 150 μm thick (Fig. 11.1b). The oxygen tension in cell suspension is controlled by exchanging the gas supply in the gas channel through the PDMS film, controlled by a programmable automatic microvalve. The electrode pads were connected to a function generator (not shown) to induce electrodeformation of cells for the biomechanical characterization of cell membranes in the cell channel. The root mean square value of the supplied voltage was 3 V at 1.58 MHz for 2 s and 0 V for 2 s, respectively. This platform was then used to measure the viscoelastic response of RBCs, under two distinct gaseous conditions, including an oxygenated condition (133 mmHg) and a deoxygenated condition (0 mmHg), by supplying the gas channel with a gas mixture (17.5% oxygen, 5% carbon dioxide, balance of nitrogen) and a gas mixture (5% carbon dioxide with the balance of nitrogen), respectively. Blood sample was obtained from a healthy donor, following the Institutional Review Board (IRB) approval from Florida Atlantic University. RBCs were washed using phosphate buffered saline buffer and resuspended to a hematocrit value of 0.1% in a low electrical conductivity (0.15 S/m) isotonic buffer, following published protocol [10].

11.3 Analysis

Cell behavior was observed under an Olympus X81 inverted optical microscope. Each cell was measured for 5 successive times under the two separate gaseous conditions. RBCs were individually tracked over time for analysis using ImageJ [13]. In present study, we report quantitative measurements of the maximum extension ratio, λ_{max} and characteristic relaxation time, t_c of individually tracked RBCs. Detailed methods for cell morphology analysis and biomechanical characterization were given in our previous publication [10]. Briefly, value of λ_{max} were evaluated by the ratio of the initial diameter and the maximum diameter in the transverse axis when cell membrane deformation reached a plateau. Value of t_c was obtained by fitting membrane extension ratio, λ as a function of time to an exponential function. Statistical analyses were performed using OriginPro 9. A paired t-test between measurements of two gaseous conditions was used to generate the p values. Effects of oxygen tension on the viscoelastic behavior of RBCs, evaluated by both parameters are significant ($p < 0.001$ and $p < 0.01$), as shown in Fig. 11.2. RBCs are more deformable under high oxygen tension condition, with the maximum λ_{max} being 1.57 ± 0.1 and dropped to 1.46 ± 0.08 under low oxygen tension. Cell deformation under both conditions follow normal distributions. The median values of t_c are 0.14 ± 0.07 s and 0.12 ± 0.06 s for the high oxygen tension and low oxygen tension measurements, respectively. The characteristic relaxation times of cell membranes under both conditions follow lognormal distributions. These results demonstrate that the influence of hypoxia on cellular viscoelastic behavior is remarkable.

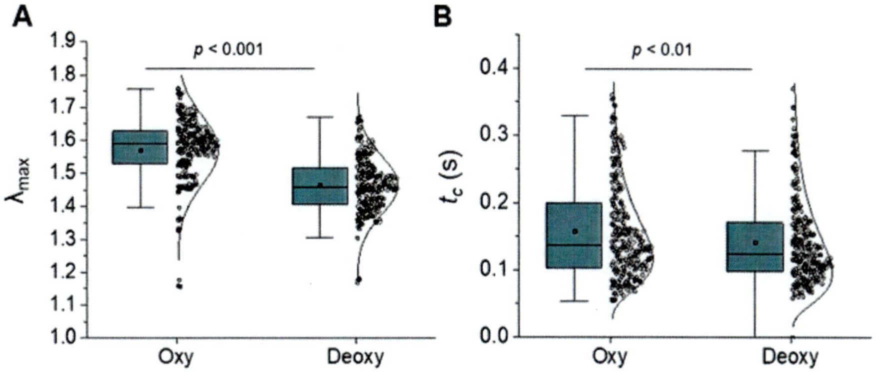

Fig. 11.2 Effects of oxygen tension on the viscoelastic response of RBCs ($n = 49$): (**a**) maximum extension ratio evaluated in the transverse axis and (**b**) characteristic relaxation time of cell membranes. Each cell was measured for successive 5 times. Each symbol represents a single cell measurement. Normal distribution and lognormal distribution are shown for λ_{max} and t_c, respectively. Oxy – 17.5% O_2, Deoxy – 0% O_2

11.4 Conclusion

This paper showed a development of microfluidics-based biomechanical testing device for cells under controlled gaseous environment. Effects of oxygen tension on the deformability and relaxation characteristics of cell membranes were obtained in normal RBCs. The stressful conditions of RBCs encountering in the complex microvasculature during blood circulation are complicated. The exact mechanisms underlying membrane fatigue failure and degradation are not yet completely understood. The developed experimental strategy can be useful to replicate some representative loading scenarios that circulating RBCs experience, paving a way to systematically study the effects of the repeated oxidative damage on cell membrane integrity and mechanics, similar to the mechanical fatigue of circulating cells [10], as well as the coupling between the mechanical fatigue and fatigue induced by cyclic hypoxia. The reported methodology can also be applied to study the interactions between oxygen tension and biomechanical signals in other types of biological cells.

Acknowledgements This work was supported by the NSF Grant No. 1635312, No. 1464102, and NIH Grant 1OT2HL152638. E.D. and Y.Q. acknowledge support from NIH Grant 5R01EB025819.

References

1. Mohandas, N., Gallagher, P.G.: Red cell membrane: past, present, and future. Blood. **112**(10), 3939–3948 (2008)
2. Lang, F., Lang, E., Foller, M.: Physiology and pathophysiology of eryptosis. Transfus. Med. Hemother. **39**(5), 308–314 (2012)
3. Suresh, S., Spatz, J., Mills, J.P., Micoulet, A., Dao, M., Lim, C.T., Beil, M., Seufferlein, T.: Connections between single-cell biomechanics and human disease states: gastrointestinal cancer and malaria. Acta Biomater. **1**(1), 15–30 (2005)
4. Suresh, S.: Biomechanics and biophysics of cancer cells. Acta Biomater. **3**(4), 413–438 (2007)
5. Discher, D., Dong, C., Fredberg, J.J., Guilak, F., Ingber, D., Janmey, P., Kamm, R.D., Schmid-Schonbein, G.W., Weinbaum, S.: Biomechanics: cell research and applications for the next decade. Ann. Biomed. Eng. **37**(5), 847–859 (2009)
6. Barabino, G.A., Platt, M.O., Kaul, D.K.: Sickle Cell Biomechanics. Annu. Rev. Biomed. Eng. **12**, 345–367 (2010)
7. Sakuma, S., Kuroda, K., Tsai, C.H.D., Fukui, W., Arai, F., Kaneko, M.: Red blood cell fatigue evaluation based on the close-encountering point between extensibility and recoverability. Lab Chip. **14**(6), 1135–1141 (2014)
8. Liu, J., Qiang, Y., Alvarez, O., Du, E.: Electrical impedance microflow cytometry with oxygen control for detection of sickle cells. Sensors Actuators B Chem. **255**, 2392–2398 (2018)
9. Du, E., Diez-Silva, M., Kato, G.J., Dao, M., Suresh, S.: Kinetics of sickle cell biorheology and implications for painful vasoocclusive crisis. Proc. Natl. Acad. Sci. **112**(5), 1422–1427 (2015)
10. Qiang, Y., Liu, J., Dao, M., Suresh, S., Du, E.: Mechanical fatigue of human red blood cells. Proc. Natl. Acad. Sci. **116**(40), 19828–19834 (2019)
11. Qiang, Y., Liu, J., Du, E.: Dynamic fatigue measurement of human erythrocytes using dielectrophoresis. Acta Biomater. **57**, 352–362 (2017)
12. Qiang, Y., Liu, J., Yang, F., Dieujuste, D., Du, E.: Modeling erythrocyte electrodeformation in response to amplitude modulated electric waveforms. Sci. Rep. **8**(1), 10224 (2018)
13. Schneider, C.A., Rasband, W.S., Eliceiri, K.W.: NIH Image to ImageJ: 25 years of image analysis. Nat. Methods. **9**(7), 671–675 (2012)

Check for updates

High speed Holographic Shape and Vibration Measurement of the Semi-transparent Tympanic Membrane

12

Haimi Tang, Pavel Psota, John J. Rosowski, Jeffery T. Cheng, and Cosme Furlong

Abstract

Holographic interferometric methods have been used by our group to quantify the shape and full-field vibrations of the tympanic membrane (TM) in the human cadaveric ear. Previously, we integrated a high-speed holographic system using a 532 nm wavelength laser for displacement measurement and a 770–789 nm tunable wavelength laser for shape measurement. However, because of the semi-transparency of the TM and the short exposure time (<13 μs) from high-speed recording (>75,000 fps), the TM surface needs to be painted with a thin layer of white paint to increase reflection of the laser beam to obtain quantifiable interference patterns. The need for painting is not practical for in-vivo live human ear measurement. In this paper, we describe our efforts to measure the unpainted semi-transparent TM. We adapted a multi-angle illumination shape measurement method into the existing holographic system to eliminate the need for the tunable laser source. It allows us to use the single wavelength (532 nm) high power laser to perform both shape and displacement measurements on the unpainted TM. We show results from three fresh postmortem human ears before and after experimentally simulated common middle-ear pathologies. The Frequency Response Function (FRF) of the displacement normalized by the sound stimulus at each pixel point of the entire TM surface was calculated. Vibration parameters such as dominant frequency, rise time, settling time over the entire surface of the TM are then determined and compared among different middle ear conditions. We observe distinguishable changes in these vibration parameters, which may be useful for the differential diagnosis of middle ear pathologies.

H. Tang (✉)
Center for Holographic Studies and Laser micro-mechaTronics—CHSLT, Worcester, MA, USA

Mechanical Engineering Department, Worcester Polytechnic Institute, Worcester, MA, USA
e-mail: HTANG3@WPI.EDU

P. Psota
Faculty of Mechatronics, Informatics and Interdisciplinary Studies, Technical University of Liberec, Liberec, Czech Republic

J. J. Rosowski · J. T. Cheng
Eaton-Peabody Laboratory, Massachusetts Eye and Ear, Boston, MA, USA

Department of Otolaryngology, Harvard Medical School, Boston, MA, USA

C. Furlong
Center for Holographic Studies and Laser micro-mechaTronics—CHSLT, Worcester, MA, USA

Mechanical Engineering Department, Worcester Polytechnic Institute, Worcester, MA, USA

Eaton-Peabody Laboratory, Massachusetts Eye and Ear, Boston, MA, USA

Department of Otolaryngology, Harvard Medical School, Boston, MA, USA

J. Notbohm et al. (eds.), *Mechanics of Biological Systems and Materials & Micro-and Nanomechanics & Research Applications*, Conference Proceedings of the Society for Experimental Mechanics Series,
https://doi.org/10.1007/978-3-030-59765-8_12

Keywords

Frequency response function, High-speed holography, Human tympanic membrane, Single wavelength laser, Shape and vibration measurement

12.1 Introduction

The acoustic-mechanical energy transformations start with Tympanic Membrane (TM) and are affected by its mechanical properties, shape, thickness, and microstructure and the load from the ossicular chain and inner ear [1]. We developed a High-speed Digital Holographic (HDH) system to measure acoustically induced transient displacements and shapes of live mammalian Tympanic Membrane to investigate TM's mechanical properties and to better understand the acoustic to solid interaction of the TM during the hearing process. However, due to the semi-transparency of the TM, its surface has to be prepared to increase reflection for high frame rate recording. In this paper, we present our latest improvements to the HDH system to overcome the need for TM surface preparation. The improved system is capable of determining the shape of the entire TM with a spatial resolution less than 150 μm using a multi-angle illumination holographic method and measurements of nanometer-scale click-induced displacements with a temporal resolution of less than 20 μs. Both shape and displacement measurements are performed with a single frequency laser resource.

12.2 Methods

Digital holography utilizes a coherent light source to illuminate the object, and it can record and reconstruct the optical wavefield (amplitude and phase) scattered from the object. The surface contour of the sample's shape can be measured by phase-shifting digital holography [2, 3]. For the experimental arrangement shown in Fig. 12.1, by introducing m steps of controlled change of the angle $d\theta$ to the illumination beam, the height (z) of the measured surface is [4]:

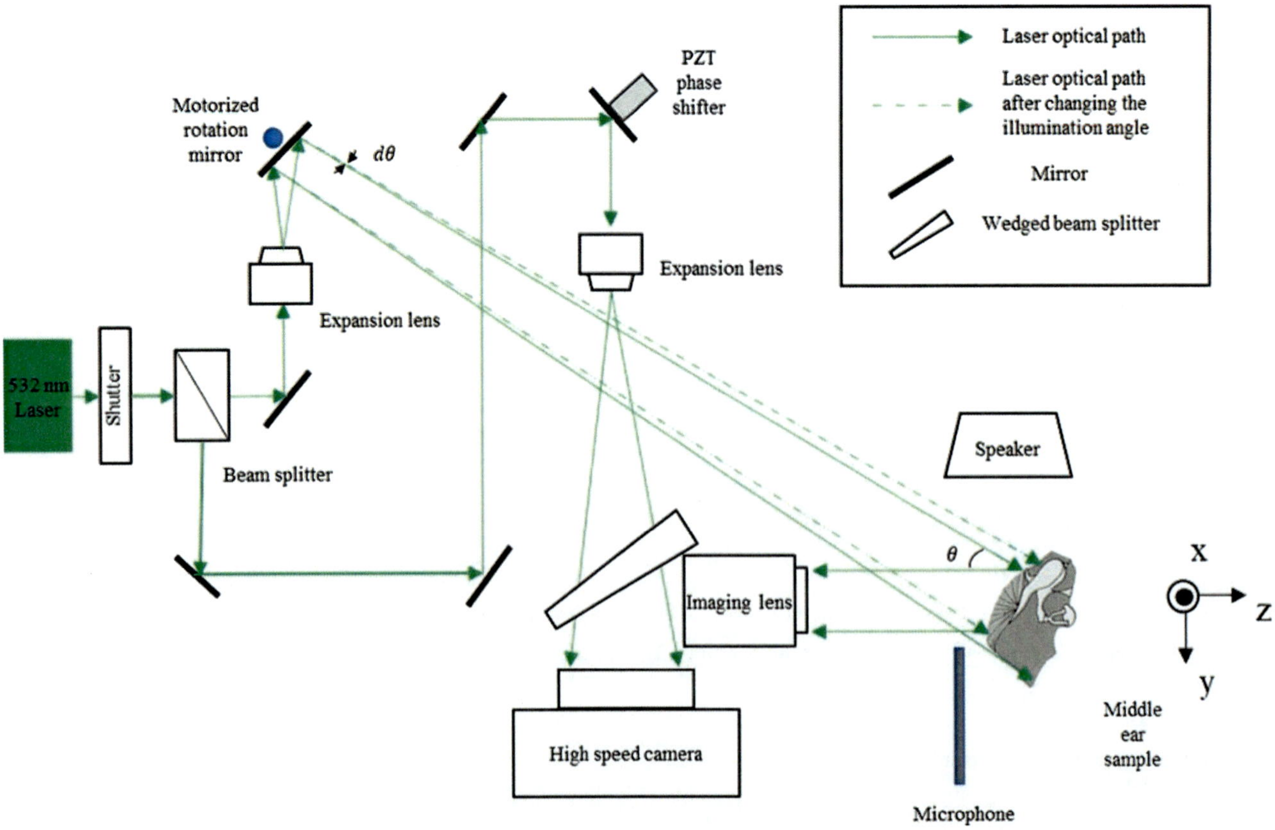

Fig. 12.1 Schematic of the HDH shape and vibration measurement system. The system utilizes a single laser resource (200 mW) to perform both the shape and vibration measurements

$$z = dP_m \left(\sum_{m=1}^{m} \frac{2\pi}{\lambda} \sin\theta \, d\theta_m \right)^{-1}, \tag{12.1}$$

where dP_m is the total optical phase difference due to the m steps changes of the illumination angle; λ is the wavelength of the laser; θ is the angle of illumination. θ is assumed to be constant when $d\theta$ is small.

The displacement measurement method utilizes Pearson's correlation to locate two undeformed frames with 0 and $\pi/2$ phase shift prior to the sample's excitation to measure the full-field-of-view (>200,000 points at a 67,200 camera frame rate) transient displacements. These two undeformed frames are used to calculate the optical phase of a deformed frame $\Delta\Phi(x, y, t)$ at time t, where (x,y) is the pixel coordinates of the sensor. From the shape measurement result, the spatial varied sensitivity vector, $\boldsymbol{K}(x,y)$, and the surface normal unit vector, $\boldsymbol{n}(x,y)$, are used to derive the surface normal displacement (s) at location (x,y) and time t [5]:

$$s(x,y,t) = \frac{\Delta\$(x,y,t)}{K(x,y) \bullet n(x,y)} \tag{12.2}$$

The schematic of the experimental setups of the digital holographic system is shown in Fig. 12.1. A single wavelength laser (532 nm) source is divided into the reference beam and the object beam by a tunable beam splitter. A piezoelectric actuated mirror is used to modulate the phase of the reference's optical path. A motorized mirror on the object path is used to change the illumination angle for shape measurement. An imaging lens is used to collect the scattered wavefront from the object. And both the object wavefront and reference wavefront are combined by the wedged beam splitter in front of the high-speed camera. A speaker is used to deliver acoustic excitation to the sample, and a high-frequency microphone is placed in close distance to the sample to monitor the acoustic single received by the sample.

To test if our high-speed holographic system is sensitive to distinguish various middle ear conditions, we designed the following experiments: first, we measured the shape and vibration of the TM when the middle ear cavity was open; next, we sealed the middle ear cavity and measured the vibration and shape of the TM again as the control measurement. We then injected sailing to the sealed middle ear cavity to fill either half of the middle ear space or full space to mimic Otitis Media with Effusion (the most common middle ear disease) and repeated our measurements. We then removed sailing and added a tiny drop of superglue to the Stapes head to mimic Stapes fixation or cut the Incudo-Stapedial (IS) joint with a myringotomy knife to mimic IS joint interruption and repeated the measurements. We recorded acoustic excitation level for each measurement with a microphone positioned near the TM surface.

The Frequency response function (FRF) at each measured point is the ratio of the fast Fourier transform (FFT) of the displacements s (response) and the FFT of sound pressure p (stimulus):

$$FRF(x,y,\omega) = \frac{FFT\left[s(x,y,t)\right]}{FFT[(p(t)]}, \tag{12.3}$$

where x and y are pixel coordinates; t is time, and ω is frequency; $p(t)$ is the acoustic signal in the time domain. Three different surface maps based on full-field TM vibrations are also computed. The dominant frequency map is defined by locating the maximum power of the FRF of each pixel in the frequency domain; the rising time map is the time for each measured point to reach its maximum displacement; the settling time map is the time for the displacement at each point to converge to zero. Rising time and settling time are based on the impulse response of each point by taking the inverse FFT of the FRF. These maps are used to visualize and compare vibration patterns of the TM with different middle ear manipulations.

12.3 Representative Results

Preliminary displacements, shape measurements, and vibration analyses were performed on a fresh cadaveric human TM without surface painting before and after various simulated middle ear pathologies. The 3D shape of the TM from each manipulation is used to compensate for the holographic sensitivity variation due to small change of the optical setup after each manipulation [6]. A representative shape measurement is shown in Fig. 12.2, when the middle ear cavity is closed. The shape measurement can achieve resolution higher than 150 micrometers [4].

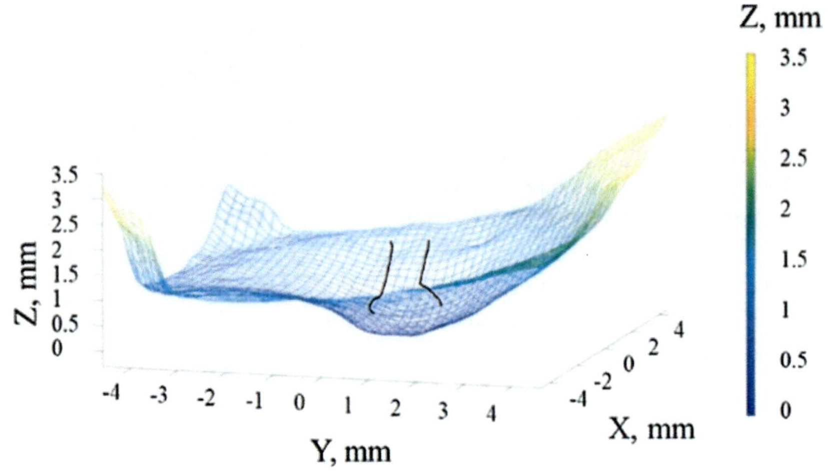

Fig. 12.2 A 3D presentation of measured TM shape. The shape is measured after closing the middle ear cavity. The black line shows the location of the manubrium and umbo

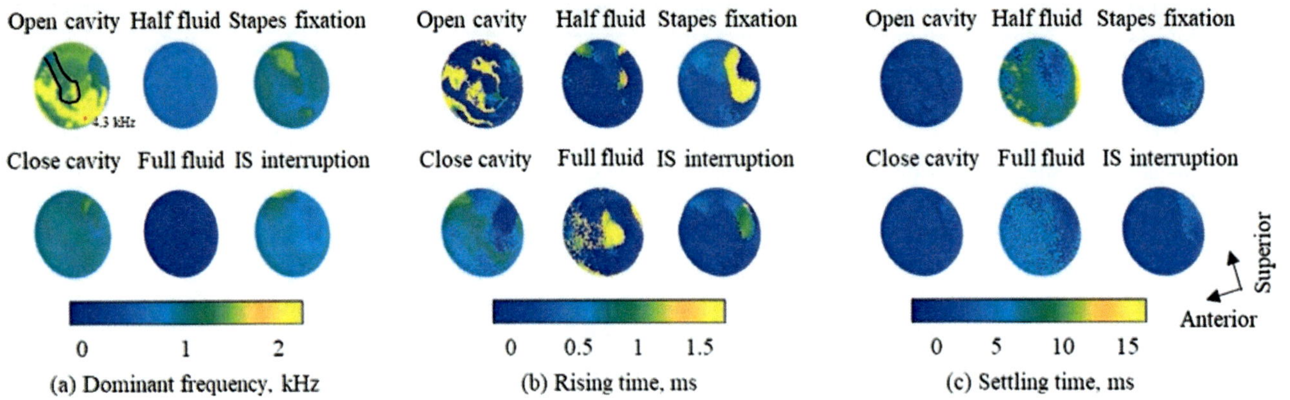

Fig. 12.3 Maps of vibration parameters for different middle ear conditions: (**a**) dominant frequency: the saturated area of the open cavity case has a dominant frequency of 4.3 kHz, (**b**) rising time, and (**c**) settling time

The dominant frequency, the rising time, and the settling time are plotted for the entire surface of the TM at six different middle ear manipulations in Fig. 12.3. The black line in the dominant frequency map with open cavity outlines the location of the manubrium and umbo. The anterior and superior directions are marked with the black arrows in the bottom right of the settling time map. The values of computed parameters on each map are coded in colors as shown in color bars in the bottom.

Considering the closed cavity as the control measurement, the half-fluid injection case has an overall reduced dominant frequency of around 0.7 kHz. The full fluid injection case has a near-zero dominant frequency because the fluid greatly reduced vibration of the TM. The Stape fixation case has a higher dominant frequency, longer rising time, and longer settling time on the posterosuperior quadrant, indicating localized changes of these parameters due to middle ear condition. The IS joint interruption showed a similar changing trend in the same area; however, the rising time and settling time are smaller compared to the Stapes fixation condition.

12.4 Conclusion

We successfully measured the semi-transparent TM shape and vibration using the high-speed holographic system with a single-wavelength laser source. We analyzed the time domain and the frequency domain displacement results of the acoustically induced TM impulse response. Those abovementioned motion parameters (dominant frequency, rising time, and settling time) provide indicators to distinguish the vibration patterns of the TM under different middle ear manipulations. Those indicators may be used for diagnostic applications in the future. We will continue our experiments to include more samples and upgrade the current system to measure live animal and human subjects.

Acknowledgments This work has been funded (R01DC016079) by the National Institute on Deafness and Other Communication Disorders (NIDCD), the National Institute of Health (NIH), and the Massachusetts Eye and Ear Infirmary (MEEI). The authors also gratefully acknowledge the support of the Mechanical Engineering Department, Worcester Polytechnic Institute, and contributions by graduate and undergraduate students of the CHSLT and MEEI.

References

1. Razavi, P., Cheng, J.T., Furlong, C., Rosowski, J.J.: High-speed holography for in-vivo measurement of acoustically induced motions of mammalian tympanic membrane. Mechanics of Biological Systems and Materials (2016)
2. Ichirou, Y., Ohta, S., Kato, J.: Surface contouring by phase-shifting digital holography. Opt. Lasers Eng. **36**(5), 417–428 (2001)
3. Soenke, S., Osten, W., Jueptner, W.. Measuring shape and deformation of small objects using digital holography. Laser Interferometry IX: Applications, vol. **3479**, pp. 104–115 (1998)
4. Psota, P., Tang, H., Pooladvand, K., Vít, L., Furlong, C., Rosowski, J.J., Cheng, J.T.: Investigation of tympanic membrane shape using digital holography. Optics and Measurement International Conference 2019, vol. **11385**, pp. 113850G (2019)
5. Tang, H., Razavi, P., Pooladvand, K., Psota, P., Maftoon, N., Rosowski, J.J., Furlong, C., Cheng, J.T.: High-speed holographic shape and full-field displacement measurements of the tympanic membrane in normal and experimentally simulated pathological ears. Appl. Sci. **9**, 2809 (2019)
6. Razavi, P., Tang, H., Rosowski, J.J., Furlong, C., Cheng, J.T.: Combined high-speed holographic shape and full-field displacement measurements of tympanic membrane. J. Biomed. Opt. **24**(3), 031008 (2018)

Adhesion Index: A Novel Biocompatibility Assessment Standard for Medical Devices

13

James D. Boyd, Craig S. Miller, and Martha E. Grady

Abstract

Tens of millions of Americans receive medical devices and procedures annually at a cost of $170 billion. Unfortunately, over half of all hospital acquired infections result from a permanent or temporary implanted device procedure. Although deleterious bacteria adherent to medically implanted devices are major contributors to this problem, biocompatibility standards lack consideration for these adherence properties. A biocompatibility assessment that considers both the adhesion of invasive bacteria and host cells is needed to advance patient care. The quantitative determination of the effect of implant surface characteristics on bacterial adhesion, as well as our bodies natural cell adhesion, will lead to the development of more successful medical implants. The laser spallation technique has been used previously for measurements of biofilm and cell adhesion. Laser spallation has distinct advantages as it results in quantified adhesion strength through a non-contact high strain rate force. The non-contact force does not disturb or alter the living films pre-testing, which ensures accurate adhesion measurements. This mechanical test is implemented alongside one dimensional wave analysis and an interface adhesion strength is measured for the biomaterial-implant mimicking surface interfaces. In this study, use of laser spallation on a titanium surface mimicking a dental implant device is explored. *Streptococcus mutans*, an oral gram-positive facultative anaerobe, is chosen because it is a common oral microbe and earlier colonizer of dental biofilms. In as much as titanium dental implants require osseous-integration, MG 63, selected for its numerous osteoblastic traits that are typical of an immature osteoblast, is optimal for initial adhesion studies. Bacteria and cells are cultured on titanium substrates that have undergone large grit sandblasting to mimic the roughness of dental implant surfaces. We demonstrate a quantitative evaluation of adhesion strength of an oral biofilm on dental implant mimicking surfaces through the adaptation of the laser spallation technique. The adhesion values for both invasive bacteria and constructive cells are compared to determine the biocompatibility for dental implant surfaces, by obtaining a unitless ratio referred to as the Adhesion Index. The Adhesion Index can be used to determine the biocompatibility of other medical implant devices and optimize future implant designs.

Keywords

Biofilms · Laser spallation, Adhesion, *Streptococcus Mutans*, MG-63, Dental implants

J. D. Boyd (✉) · M. E. Grady
Department of Mechanical Engineering, College of Engineering, University of Kentucky, Lexington, KY, USA
e-mail: jdbo227@g.uky.edu

C. S. Miller
Department of Oral Health Practice, College of Dentistry, University of Kentucky, Lexington, KY, USA

Department of Oral Health Research, College of Dentistry, University of Kentucky, Lexington, KY, USA

J. Notbohm et al. (eds.), *Mechanics of Biological Systems and Materials & Micro-and Nanomechanics & Research Applications*, Conference Proceedings of the Society for Experimental Mechanics Series,
https://doi.org/10.1007/978-3-030-59765-8_13

The focus of this study is to examine the adhesion of oral bacteria to a titanium surface. Early colonizing bacteria are the foundational microbes that promote the adhesion and growth of more pathogenic bacteria onto the newly implanted surface. The initial bacterial colonization process in the mouth is dominated by oral *Streptococcus*, which make up over 80% of the early biofilm constituents [1]. To model typical early colonizing bacteria, *Streptococcus mutans* was selected as it is associated with dental caries and promotes the adhesion of other bacteria [2]. Dental implants are susceptible to adherent bacteria simultaneous to the natural wound healing process that begins when the titanium implant is placed in bone, i.e., as osteoblasts contribute to osseointegration [3]. To model the behavior of immature osteoblast cells, MG 63 osteosarcoma cells were selected for their excellent immature osteoblastic traits [4]. To mimic a dental implant surface, the substrate system is coated with commercially pure titanium. Both smooth and roughened titanium surfaces were examined in order to mimic the varying surface roughnesses found on dental implants [5]. The dental implant mimicking substrate is inoculated with either the *S. mutans* or MG 63 cells. *S. mutans* bacteria are cultured in Todd Hewitt Yeast with sucrose, while MG 63 is cultured in Eagle's Minimum Essential Medium, until confluent biofilms and monolayers are produced respectively. The sample assembly, cultured with appropriate films, then undergo laser spallation experiments to determine the sample-biofilm interfacial adhesion strength. Adhesion strength values for both films on the smooth and rough surfaces are directly comparable values that can determine the effects of surface morphology on biocompatibility. The smooth surface titanium is used as a baseline and the effect of surface roughness is examined. In an ideal implant device, the surface would promote the adhesion of host cell formation and deter the formation of deleterious biofilms. The ratio of the adhesion of cells to the adhesion of biofilms can be used as a non-dimensional value, which will predict the biocompatible behavior of the implant, with higher values indicating favorable host response. This value, called here the Adhesion Index, provides insight into the biocompatibility and surface properties of implant materials relative to cell adhesion [6].

In order to quantify the adhesion of bacterial biofilms and cell monolayers several techniques are available. The most commonly used adhesion technique used in bacterial studies are counting methods [7]. Unfortunately counting methods while excellent at providing numerical values for bacteria present, they fail to provide quantified adhesion strength values that could be directly comparable. Jet impingement studies also have been performed to quantify the adhesion of bacterial and cellular films, however these tests rely on contact methods to induce stresses and can often lead to deformation of the films before accurate measurements can be obtained [8]. The low cohesiveness of these films can cause them to fail cohesively during testing before adhesion can be measured accurately [9].

Laser spallation is a thin film adhesion test that previously has been used to quantify the adhesion of metallic films onto substrates [10, 11]. More recently the laser spallation technique has been used to measure the adhesion of bacterial and cellular films [12–15]. The laser spallation setup used is illustrated in Fig. 13.1a. Laser spallation operates by delivering a photoacoustic shockwave to the biofilm which, with sufficient energy, will spall the films from the interface surface. To obtain the magnitude and shape of this photoacoustic shockwave, calibration experiments are performed to obtain the free surface displacement during loading using laser interferometry. This method is superior to other film adhesion testing methods as it is a non-contact method that targets only a specific portion of the sample, leaving the rest of the area unscathed and available for further experimentation. Laser spallation also will cause direct adhesive failure, not cohesive failure.

Dental implants include smooth and rough titanium surfaces, depicted in Fig. 13.1b. The threads have been roughened to increase osseointegration for years, however the effect of this roughness on bacterial adhesion is debated [7, 16, 17]. To mimic these surface characteristics glass slides are purchased with e-beam evaporated 100 nm coatings of titanium. The opposite side of the slide is coated in 300 nm of aluminum which acts as the energy absorbing layer for the laser used. A separate set of slides are first sandblasted with large grit to obtain a surface roughness between 1–1.5 μm. The glass slides are cut into 1″×1″ and the titanium surface is adhered to the bottom of a 35 mm petri dish with a 13/16″ hole, using biologically inert silicone [12]. In order to obtain powerful enough stress waves to initiate delamination, a waterglass layer was added to the back side of the samples, over the aluminum layer. Samples are then adhered to the bottom of a 35 mm diameter petri dish using a biologically inert silicone. Bacteria or cells are then inoculated inside of the substrate assembly until confluent films are formed.

Multiple substrate assemblies of biofilm and cells are tested, and multiple locations on each film are loaded. Onset of failure of the bacterial biomaterial-titanium interface is marked by spallation of the biofilm from the surface. For example, Fig. 13.1c depicts the adhesion failure progression for both *S. mutans* and MG 63 films from both smooth and rough titanium. As fluence values increase the spallation increases for all of the films tested. During testing the failure statistics for each film at each fluence are recorded in order to determine the minimal energy needed to initiate spallation. Uncultured substrates are used as calibration specimen and are loaded with identical fluences used in the spallation study, and the free surface displacement is measured using a Michelson interferometer. The substrate stress profile can be measured using the free surface velocity and converted into interface strength using previously accepted methods [18]. Weibull analysis [19, 20], common in

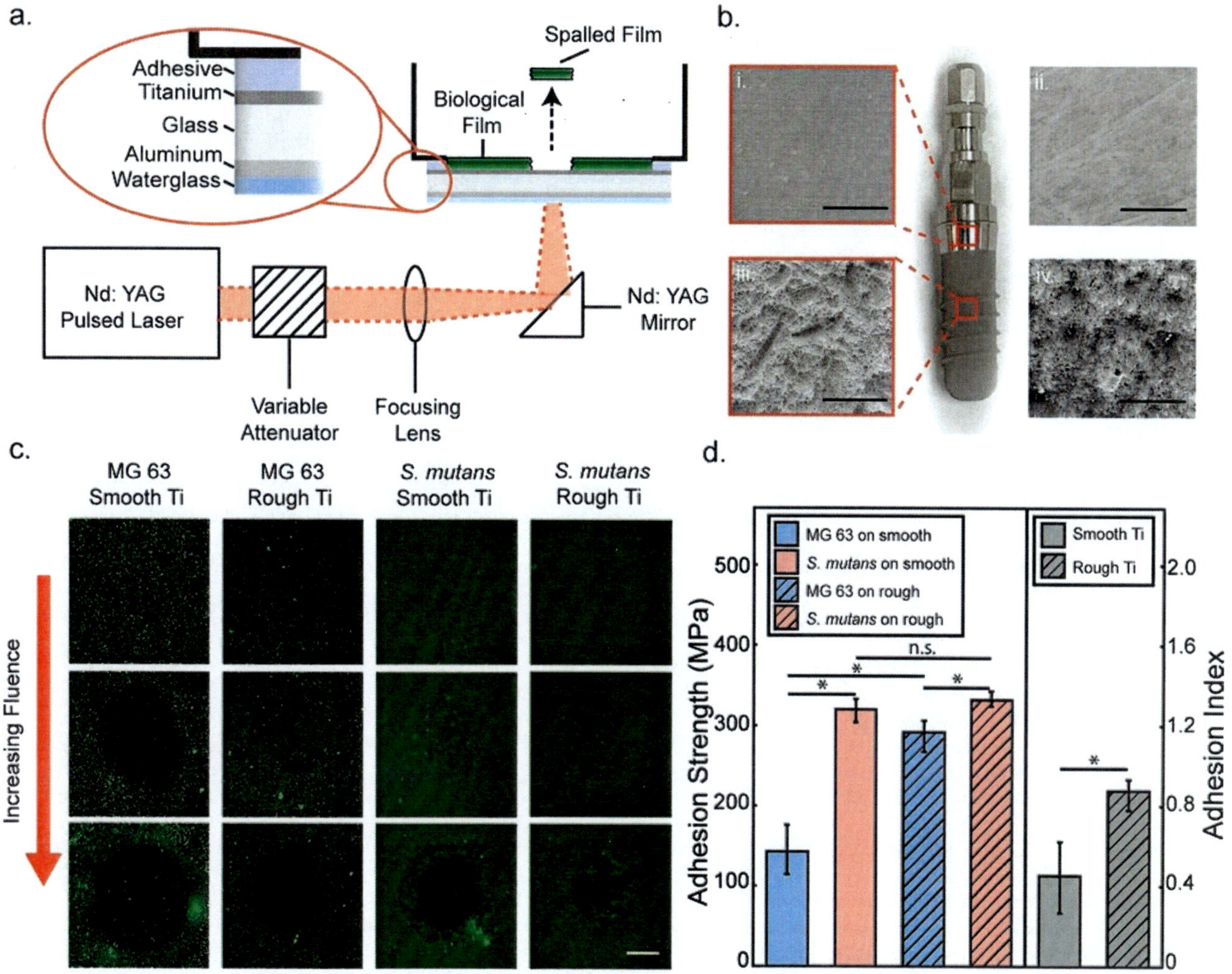

Fig. 13.1 (**a**) The laser spallation set up used during delamination experiments. The Nd: YAG laser used reflected 90° to allow for horizontal orientation of the substrate assembly used. (**b**) (1,3) Depicts SEM images obtained of indicated dental implant surfaces, (2,4) are SEM images from the test surfaces that mimic dental implants (**c**) Illustrate typical failure for both *S. mutans* and MG 63 on smooth and rough titanium surfaces due to increasing fluence values. (**d**) Graph of interface strength obtained after calibration experiments and Weibull analysis, alongside corresponding Adhesion Index value for smooth and rough titanium.

macroscopic adhesion analyses, calculates the half-life from a Weibull distribution, which is used as the adhesion strength. There is a two-fold difference in adhesion strength between MG 63 monolayers and *S. mutans* biofilms on smooth titanium. MG 63 has an adhesion strength of only 143 MPa, with a 95% confidence interval of (114, 176), and *S. mutans* has an adhesion strength of 320, with a 95% C.I. of (304,333). The adhesion strength for MG 63 increases to 292 MPa, with a 95% C.I. (267, 306), and *S. mutans* adhesion strength minimally increases to 320 MPa, with a 95% C.I. (304, 333). A ratio of the adhesion strength for bacteria and cells onto smooth and rough titanium surfaces is calculated to obtain the Adhesion Index for both films as one property of biocompatibility. When bacteria and cells are cultured onto smooth substrates the Adhesion Index is measured at 0.451, with a 95% C.I. (0.267, 0.622), but when they are cultured onto roughened surfaces the Adhesion Index increases to 0.876, with a 95% C.I. (0.780, 0.932). Adhesion strengths and Adhesion Index values are shown in Fig. 13.1d. The adhesion strength for bacterial biofilms is greater than that of the cell monolayers. However, the increase in surface roughness drastically increases the adhesion for cells while only marginally increasing adhesion for bacteria. When these values are compared it is evident that the macroscopic surface roughness plays a greater role in cell adhesion compared to bacterial adhesion resulting in an increase in Adhesion Index and biocompatibility of the surface.

This study directly compares the adhesion for *S. mutans* biofilms and MG 63 monolayers onto both smooth and roughened titanium dental implant mimicking surfaces. Laser spallation was used in order to quantify the adhesion strength for the biological films. The adhesion values were directly compared to determine the effect of surface modifications on their respective adhesion and thus the biocompatibility for the implanted device. An increase in surface roughness resulted in an increase adhesion strength for MG 63 with minimal change in *S. mutans* indicating a greater biocompatibility for these surfaces. Future work intends to further model *in vivo* conditions. Swapping out culture media with blood plasma and proteins will more closely mimic the initial wound healing process that occurs after implantation. Also, diversifying the bacteria, cells and substrate materials will allow for the testing of a variety of permanent and temporary medically implanted devices. The goal of this research is to develop an adhesion testing protocol, and metric, to aid in the design of medical devices, and ultimately be used to reduce biofilm-related infections while promoting the successful integration of these medical devices.

Acknowledgements We thank the Center for Pharmaceutical Research and Innovation (CPRI) for use of bacterial culture equipment. CPRI is supported, in part, by the University of Kentucky College of Pharmacy and Center for Clinical and Translational Science (UL1TR001998). We thank Drs. Larissa Ponomareva and Natalia Korotkova for sharing their bacterial culture expertise.

References

1. Kreth, J., Merritt, J., Qi, F.: Bacterial and host interactions of oral streptococci. DNA Cell Biol. **28**(8), 397–403 (2009)
2. Barbosa, J.O., et al.: Streptococcus mutans can modulate biofilm formation and attenuate the virulence of Candida albicans. PLoS One. **11**(3), e0150457 (2016)
3. Chang, Y.-Y., et al.: Biological characteristics of the MG-63 human osteosarcoma cells on composite tantalum carbide/amorphous carbon films. PLoS One. **9**(4), –e95590 (2014)
4. Shapira, L., Halabi, A.: Behavior of two osteoblast-like cell lines cultured on machined or rough titanium surfaces. Clin. Oral Implants Res. **20**(1), 50–55 (2009)
5. Gaviria, L., et al.: Current trends in dental implants. J. Korean Assoc. Oral Maxillofac. Surg. **40**(2), 50–60 (2014)
6. Albrektsson, T., et al.: The long-term efficacy of currently used dental implants: a review and proposed criteria of success. Int. J. Oral Maxillofac. Implants. **1**, 11–25 (1986)
7. Duarte, P.M., et al.: Bacterial adhesion on smooth and rough titanium surfaces after treatment with different instruments. J. Periodontol. **80**(11), 1824–1832 (2009)
8. Bayoudh, S., et al.: Assessing bacterial adhesion using DLVO and XDLVO theories and the jet impingement technique. Colloids Surf. B: Biointerfaces. **73**(1), 1–9 (2009)
9. Stoodley, P., et al.: Biofilm material properties as related to shear-induced deformation and detachment phenomena. J. Ind. Microbiol. Biotechnol. **29**(6), 361–367 (2002)
10. Grady, M.E., et al.: Molecular tailoring of interfacial failure. Langmuir. **30**(37), 11096–11102 (2014)
11. Wang, J.L., Weaver, R.L., Sottos, N.R.: A parametric study of laser induced thin film spallation. Exp. Mech. **42**(1), 74–83 (2002)
12. Boyd, J.D., Korotkova, N., Grady, M.E.: Adhesion of biofilms on titanium measured by laser-induced spallation. Exp. Mech. **59**(9), 1275–1284 (2019)
13. Boyd, J.D., Miller, C.S., Grady, M.E.: Biofilm and cell adhesion strength on titanium via the laser spallation technique. Mater. Data Facil. (2019)
14. Kearns, K.L., Boyd, J.D., Grady, M.E.: Biofilm rupture by laser-induced stress waves increases with loading amplitude, independent of location. ACS Appl. Bio Mater. **3**(3), 1426–1433 (2020)
15. Hu, L., et al.: Cell adhesion measurement by laser-induced stress waves. J. Appl. Phys. **100**(8), 084701 (2006)
16. Mehl, C., et al.: Adhesion of living cells to abutment materials, dentin, and adhesive luting cement with different surface qualities. Dent. Mater. **32**(12), 1524–1535 (2016)
17. Lincks, J., et al.: Response of MG63 osteoblast-like cells to titanium and titanium alloy is dependent on surface roughness and composition. Biomaterials. **19**(23), 2219–2232 (1998)
18. Kandula, S.S.V., et al.: Adhesion strength measurement of polymer dielectric interfaces using laser spallation technique. Thin Solid Films. **516**(21), 7627–7635 (2008)
19. Mei, L., et al.: Influence of surface roughness on streptococcal adhesion forces to composite resins. Dent. Mater. **27**(8), 770–778 (2011)
20. van der Mei, H.C., de Vries, J., Busscher, H.J.: Weibull analyses of bacterial interaction forces measured using AFM. Colloids Surf. B: Biointerfaces. **78**(2), 372–375 (2010)

Check for updates

Evaluate the Fidelity of Synthetic Tissues Used in Escharotomy Simulators

14

Sangrock Lee, Rahul, Vanessa Hannay, Samara Gallagher, Hanna Lee, Aisha Imtiaz, Melih Turkseven, Adam Ryason, Basiel Makled, Conner Parsey, Jack Norfleet, and Suvranu De

Abstract

Patients with full thickness circumferential burns may require an urgent and relatively high-risk procedure known as an "escharotomy" to preserve limb perfusion and sometimes airway ventilation. The burned skin is incised down to the subcutaneous level, providing decompression through the incision sites. A delay in performing an escharotomy, or an inadequately performed escharotomy, may lead to true compartment syndrome endangering life and limb. With the Army's emerging doctrine of prolonged field care, non-expert providers must be trained to perform escharotomies when indicated. However, existing physical simulations fall short of delivering the fidelity necessary to train the procedure by not providing adequate haptic and visual cues. To improve the fidelity of escharotomy simulations, gaps in performance between synthetic tissue and full thickness burned tissue must be measured. This work evaluates the fidelity of synthetic tissues used in escharotomy simulators with that of the full thickness burned tissues characterized using an *ex vivo* porcine model. Incision and cutting experiments are performed at the rates relevant to the escharotomy procedure. A statistical analysis shows a significant difference in the critical force to rupture synthetic tissues and burned porcine skin tissues.

Keywords

Escharotomy, Medical simulator, Synthetic tissue, Incision and cutting, Statistical analysis

S. Lee · A. Ryason
Department of Mechanical, Aerospace, and Nuclear Engineering, Rensselaer Polytechnic Institute, Troy, NY, USA

Center for Modeling, Simulation and Imaging in Medicine, Rensselaer Polytechnic Institute, Troy, NY, USA

Rahul (✉) · M. Turkseven
Center for Modeling, Simulation and Imaging in Medicine, Rensselaer Polytechnic Institute, Troy, NY, USA
e-mail: rahul@rpi.edu

V. Hannay
Department of Biomedical Engineering, Rensselaer Polytechnic Institute, Troy, NY, USA

S. Gallagher · H. Lee · A. Imtiaz
Department of Mechanical, Aerospace, and Nuclear Engineering, Rensselaer Polytechnic Institute, Troy, NY, USA

B. Makled · C. Parsey · J. Norfleet
U.S. Army Futures Command, Combat Capabilities Development Command Soldier Center STTC, Orlando, FL, USA

S. De
Department of Mechanical, Aerospace, and Nuclear Engineering, Rensselaer Polytechnic Institute, Troy, NY, USA

Center for Modeling, Simulation and Imaging in Medicine, Rensselaer Polytechnic Institute, Troy, NY, USA

Department of Biomedical Engineering, Rensselaer Polytechnic Institute, Troy, NY, USA

J. Notbohm et al. (eds.), *Mechanics of Biological Systems and Materials & Micro-and Nanomechanics & Research Applications*, Conference Proceedings of the Society for Experimental Mechanics Series,
https://doi.org/10.1007/978-3-030-59765-8_14

14.1 Introduction

Burns are one of the most common injuries in both civilian and combat scenarios. Acute burn injury occurs in approximately 5–20% of combat casualties [1]. The circumferential full thickness burns of the extremities and torso are prone to the compressive effect of eschar. Fluid resuscitation can rapidly exacerbate such compression through increased tissue edema, with resultant vascular obstruction, ischemia, and limb loss [2]. Early escharotomy has been recommended for patients with circumferential, deep burns involving the extremities [3–5]. Escharotomy is considered to be a relatively urgent procedure, in which the burned skin is incised down to the subcutaneous level, providing decompression through the incision sites [2]. A delay in performing a clinically indicated escharotomy, or an inadequately performed escharotomy, may lead to a true compartment syndrome mandating fasciotomy [6]. Circumferential burns involving the chest can impair ventilation and/or oxygenation due to reduced thoracic excursion. Thoracic escharotomy, performed at the bedside, can be a lifesaving intervention. The Army's medical doctrine is evolving to include prolonged field care and with burns being a common injury among soldiers, non-expert providers must be adequately trained to perform escharotomies as and when situation demands. Existing physical simulations, however, fall short in providing adequate haptic and visual cues [7]. Hence, to develop a high fidelity escharotomy simulator, it is imperative to evaluate the mechanical response of synthetic tissues used in physical simulators with respect to that of the full thickness burned tissue model.

In this work, we measure the mechanical response of synthetic phantoms and *ex vivo* full thickness burned porcine skin tissues during incision and cutting experiments. The reaction forces are measured for various cutting rates relevant to escharotomy procedures. A nonlinear Fisher discriminant analysis (FDA) [8] is performed using critical force to rupture data to evaluate the fidelity of synthetic tissues.

14.2 Methods and Materials

A custom-made experimental setup is developed to perform robotic incision and cutting experiments. A 3D-printed fixture is designed to house a scalpel at the end effector of 6-DOF robotic arm (UR5, Universal Robots) [9]. The scalpel is instrumented with a 6-DOF force/torque (F/T) sensor (Nano17, ATI-Industrial) (Fig. 14.1), which provides real-time load data at 1000 Hz with a resolution of 0.0125 N. The tissue sample was mounted on top of a balloon using a custom-made fixture. The balloon was inflated at 30 mm-Hg pressure to mimic the critical capillary closure pressure, which may mandate escharotomy to prevent irreversible tissue damage. The *ex vivo* porcine skin tissues were burned at 450 °F for 30s to induce full thickness burn following protocol developed elsewhere [10]. Experiments were performed at three cutting rates: 2 mm/s, 4 mm/s, and 8 mm/s. For each cutting rate, 15 samples of synthetic tissue (cut suit neck skin, Strategic Operations Inc.) and equal numbers of burned skin samples were tested and reaction forces were recorded. A kernel Fisher discriminant analysis (KFDA) is performed to test the null hypothesis, H_0: synthetic tissue has the same mechanical property as the full thickness burned tis-

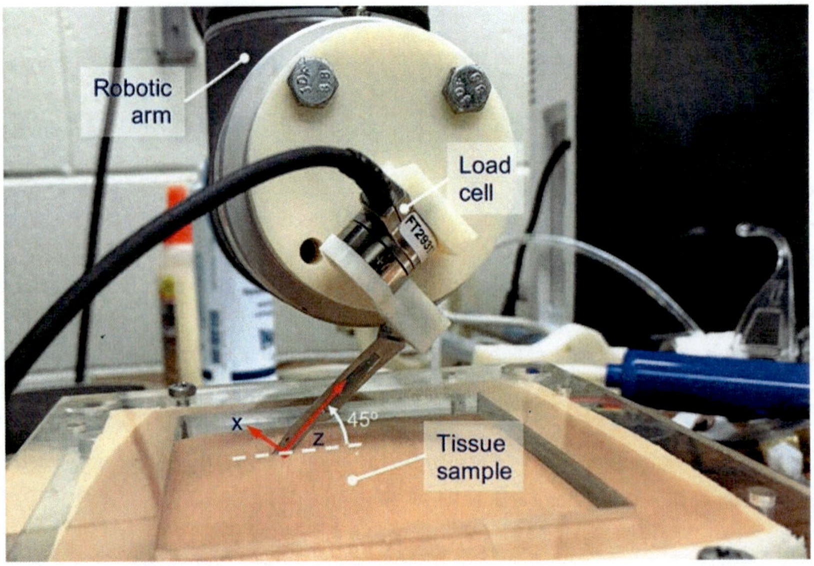

Fig. 14.1 Robotic incision and cutting experimental setup

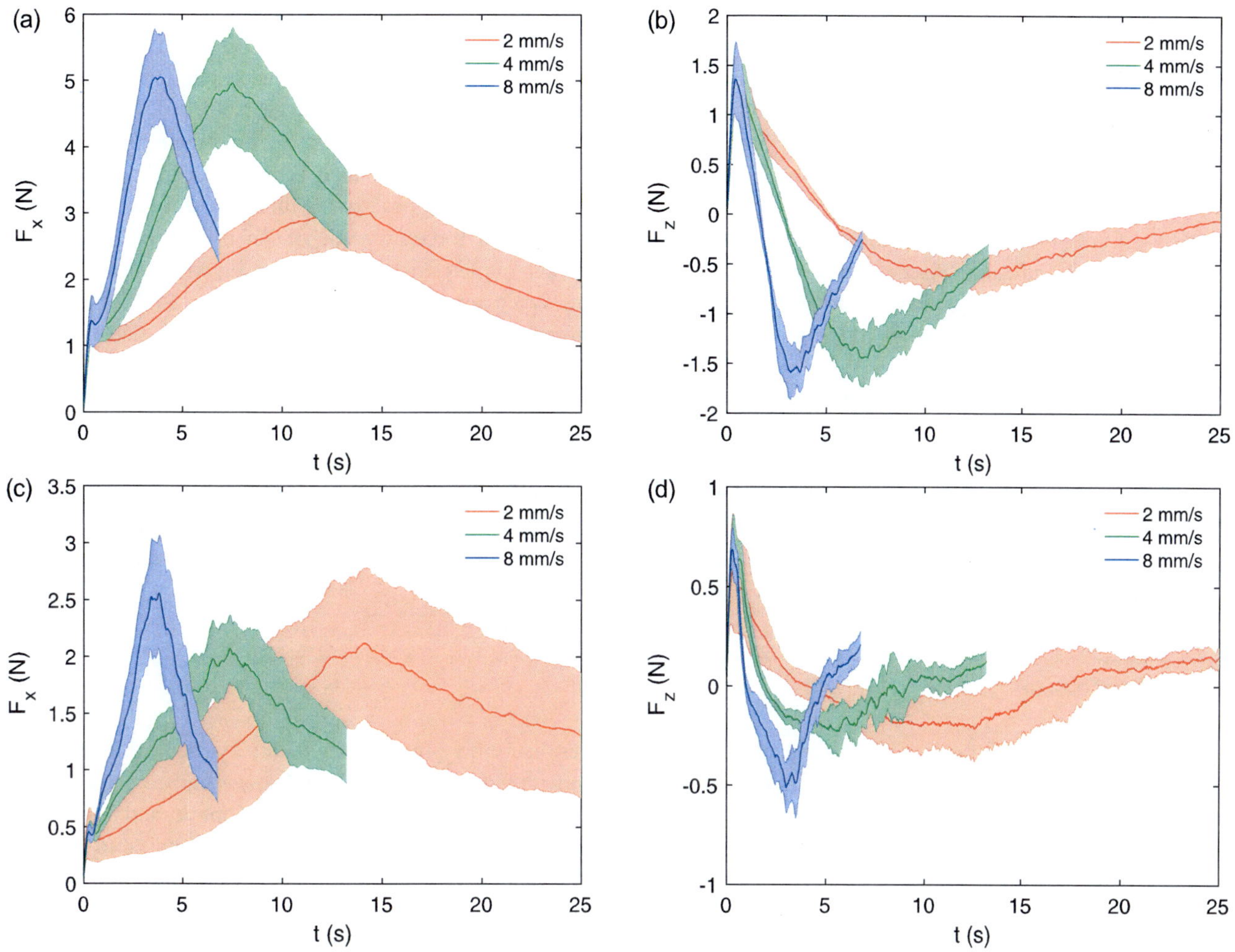

Fig. 14.2 Time-coded reaction force measured during incision and cutting experiments on (**a**, **b**) synthetic tissues and (**c**, **d**) *ex vivo* full thickness burned porcine skin tissues. Experiments are performed at three different cutting rates: 2 mm/s, 4 mm/s, and 8 mm/s

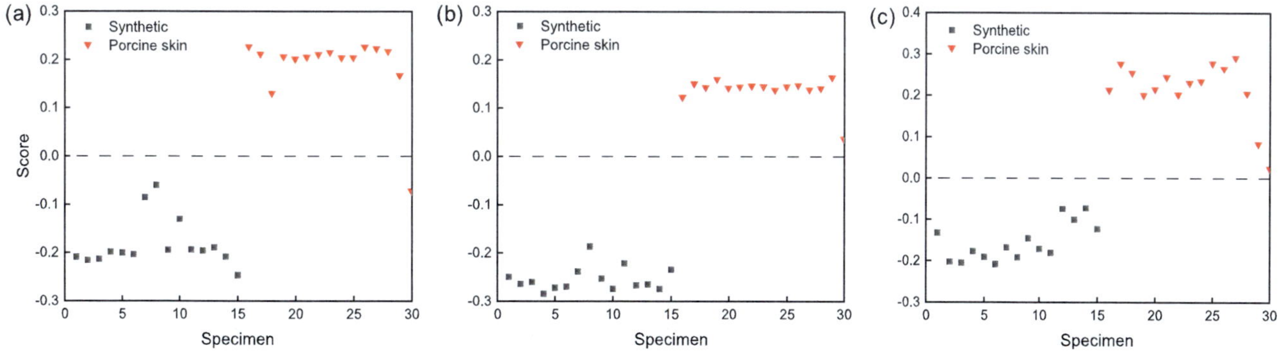

Fig. 14.3 KFDA scores for classification of synthetic vs. burned skin samples based on critical force to rupture, measured during incision and cutting experiments performed at (**a**) 2 mm/s, (**b**) 4 mm/s, and (**c**) 8 mm/s

sue, against the alternative hypothesis, H_a: synthetic and full thickness burned tissues have different mechanical properties. The KFDA is performed using measured critical force to rupture synthetic and burned skin samples. The accuracy of classification is cross-validated in a leave-one-out (LOO) fashion [11].

14.3 Results and Discussion

We begin with the comparison of reaction forces measured during incision and cutting experiments on synthetic tissues and burned skin samples. Figure 14.2 compares the reaction forces along x- and z-axis as depicted in Fig. 14.1. The force response clearly exhibits nonlinear behavior for both synthetic and burned skin tissues. Two distinct deformation regimes are observed in the time-coded force data (Fig. 14.2). During indentation, the reaction force increases sharply, indicating tissue deformation without rupture (Fig. 14.2b, d). The point of incision is recorded as a peak in the time-coded force curve, followed by a decrease of the force response. Subsequent to incision, the reaction force shows dependence on the cutting velocity (Fig. 14.2b, d), indicating viscoelastic behavior of both synthetic and burned skin tissues. However, we observe that the critical force to rupture is significantly larger for the synthetic tissues when compared to burned skin (Fig. 14.2). The mean of cutting force is much larger for synthetic tissues in comparison to burned skin tissues. These trends are consistent across all the cutting rates.

A KFDA is performed to assess the difference in mechanical response of synthetic and burned skin tissues based on cutting force to rupture data. Figure 14.3 reports the FDA scores for the pair-wise binary classification of burned skin vs. synthetic tissues for three cutting rates: 2 mm/s, 4 mm/s and 8 mm/s. The FDA score effectively provides a single defining parameter, which can then be used to compare synthetic and burned skin tissues in a pair-wise binary classification. The KFDA scores show that only one burned skin sample for 2 mm/s cutting rate case is misclassified as synthetic tissue (Fig. 14.3a). Hence, overwhelmingly rejecting the null hypothesis H_0 in favor of alternate hypothesis H_a, with the exception of one burned skin sample.

14.4 Conclusion

In this work, we evaluate the fidelity of synthetic tissues used in escharotomy physical simulators by measuring reaction forces during incision and cutting experiments and comparing them with those of *ex vivo* full thickness burned porcine skin tissues. Experiments are performed at cutting rates relevant to escharotomy simulators, *i.e.* 2 mm/s, 4 mm/s, and 8 mm/s. A nonlinear kernel Fisher discriminant analysis shows a significant difference between the critical force required to rupture synthetic tissues and burned skin tissues for all cutting rates. The synthetic tissues require much larger incision and cutting forces when compared to the burned skin tissues. The results indicate a clear mismatch between the mechanical response of burned skin tissue and that of synthetic tissues used in medical trainers. However, most conclusive evidence of the fidelity of synthetic tissues will have to come from a comparison with *in vivo* human tissues. Nonetheless, it is expected that the results of this study will provide guidance for the development of more realistic simulators for escharotomy.

Acknowledgements The authors gratefully acknowledge the support of this work through the U.S. Army Futures Command, Combat Capabilities Development Command Soldier Center STTC cooperative research agreement W911NF-17-2-0022.

References

1. Champion, H.R., Bellamy, R.F., Roberts, C.P., Leppaniemi, A.: A profile of combat injury. J. Trauma Acute Care Surg. **54**(5), S13–S19 (2003)
2. Orgill, D.P., Piccolo, N.: Escharotomy and decompressive therapies in burns. J. Burn Care Res. **30**(5), 759–768 (2009)
3. Lenhart, M. K., Savitsky, E., Eastridge, B.: Combat casualty care: lessons learned from OEF and OIF. (2012)
4. Renz, E.M., Cancio, L.C., Barillo, D.J., White, C.E., Albrecht, M.C., Thompson, C.K., Ennis, J.L., Wanek, S.M., King, J.A., Chung, K.K., Wolf, S.E., Holcomb, J.B.: Long range transport of war-related burn casualties. J. Trauma Acute Care Surg. **64**(2), S136–S145 (2008)
5. Wolf, S.E., Kauvar, D.S., Wade, C.E., Cancio, L.C., Renz, E.M., Horvath, E.E., White, C.E., Park, M.S., Wanek, S., Albrecht, M.A., Blackbourne, L.H., Barillo, D.J., Holcomb, J.B.: Comparison between civilian burns and combat burns from operation iraqi freedom and operation enduring freedom. Ann. Surg. **243**(6), 786–795 (2006)
6. Brown, R.L., Greenhalgh, D.G., Kagan, R.J., Warden, G.D.: The adequacy of limb escharotomies-fasciotomies after referral to a major burn center. J. Trauma Acute Care Surg. **37**(6), 916–920 (1994)
7. Norfleet, J., Mazzeo, M., Palata, K., Tenorio, L. M., Barocas, V., Sweet, R.: Thoracostomy simulations: a comparison of the mechanical properties of human pleura vs synthetic training pleura. MODSIM World (2016)
8. Mika, S., Ratsch, G., Weston, J., Scholkopf, B., Mullers, K.R.: Fisher discriminant analysis with kernels. In: Neural Networks for Signal Processing IX: Proceedings of the 1999 IEEE Signal Processing Society Workshop, pp. 41–48 (1999)
9. Ryason, A., Sankaranarayanan, G., Butler, K. L., DeMoya, M., and De, S.: 3D force/torque characterization of emergency cricothyroidotomy procedure using an instrumented scalpel. In: Proceedings of the IEEE 38th Annual Engineering in Medicine and Biology Society (EMBS) Conference, pp. 2145–2148 (2016)
10. Ye, H., Rahul, Dargar, S., Kruger, U., De, S.: Ultrasound elastography reliably identifies altered mechanical properties of burned soft tissues. Burns. **44**(6), 1521–1530 (2018)
11. James, G., Witten, D., Hastie, T., Tibshirani, R.: An introduction to statistical learning, vol. 112, pp. 3–7. Springer, New York (2013)

Bacterial Cell Wall Glycopolymers Affect Polymer Chain Alignment and Mechanics of *Streptococcus mutans*

15

Joree N. Sandin, Natalia Korotkova, and Martha E. Grady

Abstract

Antibiotic resistance is the phenomenon in which bacteria is or becomes resistant to antibiotics. As it pertains to oral bacteria, current methods for elimination of their presence and reduction of infection is through mechanical means, such as brushing, flossing, or swishing with an alcoholic rinse. These methods loosen the bacteria from oral surfaces or attack bacteria through cell disruption by protein denaturation and lipid membrane dissolution. One strain of oral bacteria that can cause plaque accumulation, chewing difficulty, impaired speech, and dental caries is *Streptococcus mutans* (*S. mutans*). *S. mutans* is a group of Gram-positive, facultatively anaerobic, cocci bacteria that can contribute to a multitude of oral hygiene concerns, including loss of dental implants in cases of high concentration. Research suggests that modification of the cell wall may lead to new developments on how to regulate cell permeation, thereby, regulation of microbial resistance mechanisms. In this work, three *S. mutans* stains with altered peptidoglycan layers are studied through atomic force and confocal microscopy. Surface roughness differences due to molecular glycerol phosphate alterations are quantified and compared at two positions: near the pole of the cell and within the septal region. We show AFM images of bacteria scans, including the cell wall, and quantitative surface roughness measurements at the two difference regions for each strain investigated.

Keywords

Atomic force microscopy, Confocal microscopy, Surface roughness, *Streptococcus mutans*, Bacteria

Glycerol phosphates, which are structurally related sugar polymers and forms of teichoic acids, are found in the peptidoglycan layer of most Gram-positive bacterial cell walls [1]. In observational studies on bacteria, glycerol phosphates are required for normal cellular activity, such as bacterial growth and colonization, but can be removed with little degradation to the entirety of the original polymer chain [2–4]. For a complete study of various strains of *S. mutans* bacteria, a wildtype form was first selected. To compare, a mutant type containing the glycerol phosphate modification and a complemented form, which acts as a genetic control, were chosen [5]. The glycerol phosphate modification is expected to change the biosynthesis pathways of the bacteria. This change affects the wall teichoic acid layer as well as the peptidoglycan structure, therefore, affecting the cell wall mechanics of the mutant and complemented forms. This study will seek out strain-specific morphology and surface roughness for characterization.

J. N. Sandin (✉) · M. E. Grady
Department of Mechanical Engineering, University of Kentucky, Lexington, KY, USA
e-mail: joree.sandin@uky.edu

N. Korotkova
Department of Molecular and Cellular Biochemistry, University of Kentucky, Lexington, KY, USA

J. Notbohm et al. (eds.), *Mechanics of Biological Systems and Materials & Micro-and Nanomechanics & Research Applications*, Conference Proceedings of the Society for Experimental Mechanics Series,
https://doi.org/10.1007/978-3-030-59765-8_15

Usage of light microscopy and atomic force microscopy (AFM) has proved to be a useful tool to observe and study cell wall mechanics of biological materials, such as bacteria and cells [6–9]. Using AFM in conjunction with laser scanning confocal microscopy, surface roughness values and images of singe bacteria samples were successfully obtained. This protocol involves bacteria cell culture, sample immobilization and staining, concluding with imaging and data processing. Three strains of *S. mutans*, wildtype (WT), mutant (MT), and complemented (Comp), were grown overnight in Todd Hewitt Yeast (THY) with the addition of antibiotics for all cultured strains. AFM appropriate dishes (World Precision Instruments (WPI) 35 mm dish diameter, 23 mm glass diameter, 0.17 mm glass thickness, FluoroDish™ Tissue Culture with Glass Bottom dishes) were coated with poly-l-lysine solution; enough to cover the glass substrate. These dishes were selected from an approved dishes list for the JPK-Bruker Nanowizard® 4a AFM. The dishes were set for one hour at room temperature inside of a biosafety cabinet and then the poly-l-lysine solution was removed by rinsing each dish three times with phosphate buffered saline (PBS). The sample dishes were left to dry for one hour at room temperature inside of a biosafety cabinet. After overnight incubation of the *S. mutans* strains, 30 mL of THY was inoculated with 3 mL of bacterial suspension. Using a spectrophotometer, the optical density of the solution was measured, and THY levels were adjusted until the culture reached an optical density of 0.6–0.7. 1 mL of the freshly made suspended solution was transferred onto the poly-l-lysine treated petri-dishes with the addition of 1 mL of 4% paraformaldehyde, for gentle fixation. The combination of immobilization and gentle fixation techniques has shown to aid when collecting indentation and image data, via AFM, to be processed and applied to various models [10–12]. The samples were incubated in a biosafety cabinet at room temperature for 45 min. Following incubation, 5 μL of Vancomycin BODIPY™ FL Conjugate (100 μg/mL) was placed in the sample solution and dishes were incubated at room temperature inside of a biosafety cabinet for 15 minutes. After the final incubation step, the excess liquid in each dish was aspirated off and discarded. The dishes were gently washed three times with PBS and transferred to the AFM with all liquid aspirated out. The procedure is designed to yield two dishes of each strain for a total of six samples. Before employing AFM and confocal techniques, a sample stage cell dish heater is setup and utilized to keep the bacterial samples at 37 °C.

We began by loading a sample into the Nikon A1 laser scanning confocal microscope sample stage utilizing a 20x objective to check for bacteria presence. The glass bottom is located with the objective, using either difference interference contrast (DIC, brightfield) or epifluorescence. When looking in epifluorescence, the FITC filter cube is used as the Vancomycin stain fluoresces green which stained the cell wall of the *S. mutans*'. This step, using confocal microscopy in the procedure, is to confirm the presence of the bacteria before employing AFM.

After confirming bacteria presence, the Petri dish is removed from the stage, filled half-full of PBS, and returned. A cantilever tip is loaded into the JPK-Bruker Nanowizard® 4a AFM head. When using Quantitative Imaging™ mode (QI™ mode), specific to the JPK-Bruker system, on *S. mutans*, Mikromasch tips HQ:CSC17/Cr-Au BS tips are employed, with tip radius of 8 nm, force constant of 0.18 N/m, and resonance frequency of 13 kHz to complete bacteria AFM imaging and data collection. Once the tip has been carefully and safely loaded into the instrument, it is lowered down using z-stepper motors until it is centered and sharp within the focal plane. Then, the JPK laser is turned on and positioned to aim at the reflective side of the AFM cantilever where the AFM tip has been manufactured. Once this positioning is set, the tip is retracted, and the Nikon A1 100× oil immersion objective is engaged. Vacuum grease is dabbed along the edges of the cell dish heater sample ring and microscope immersion oil is carefully placed on the objective eye. The sample is returned to the instrument by placing back into the cell dish heater stage in which it is stabilized by the grease. The 100× objective is raised up until the oil wicks over the objective eye and is in contact with the WPI glass substrate. The AFM head containing the tip is placed back onto the stage and z-stepper motors are returned to a height where the tip is sharpened in the field of view, as before. The calibration of the AFM tip involves a series of careful steps. For this particular experiment, the non-contact JPK calibration with a rectangular tip is applied. Certain factors are selected that pertain to the environment being worked in, such as a water medium, specific cantilever width and length, and expected sample height, as a few examples. Once the calibration procedure is complete, the spring constant and sensitivity is recorded for future data analysis at which the sample is ready to be probed by the instrument. As an additional step, the JPK-Bruker AFM contains a Direct Digital Overlay accessory that pairs the confocal objective with the AFM head to scan over specified regions in the sample dish. Working at such a small size scale has its limitations, but the overlay feature aids in selecting more accurate regions of the substrate that contain bacteria and not just plain glass. Images produced from this technique are shown in Fig. 15.1 (B.–D.). The brightfield image in (B.) identifies a region that was scanned by the AFM found in (C.) and (D.) and then used for surface roughness analysis.

Our specific procedure cycles through a series of scans that include a field of view, 5 μm by 5 μm containing multiple bacteria, a second field of view, 1.5 μm by 1.5 μm containing a single bacterium, and a third, 300 nm by 300 nm on top of the respective bacterium. This procedure is repeated in three to five different locations in each sample dish. An example showing the scale of the experiment as well as the progression of the surface profile scans is seen in Fig. 15.1 (B.–D.).

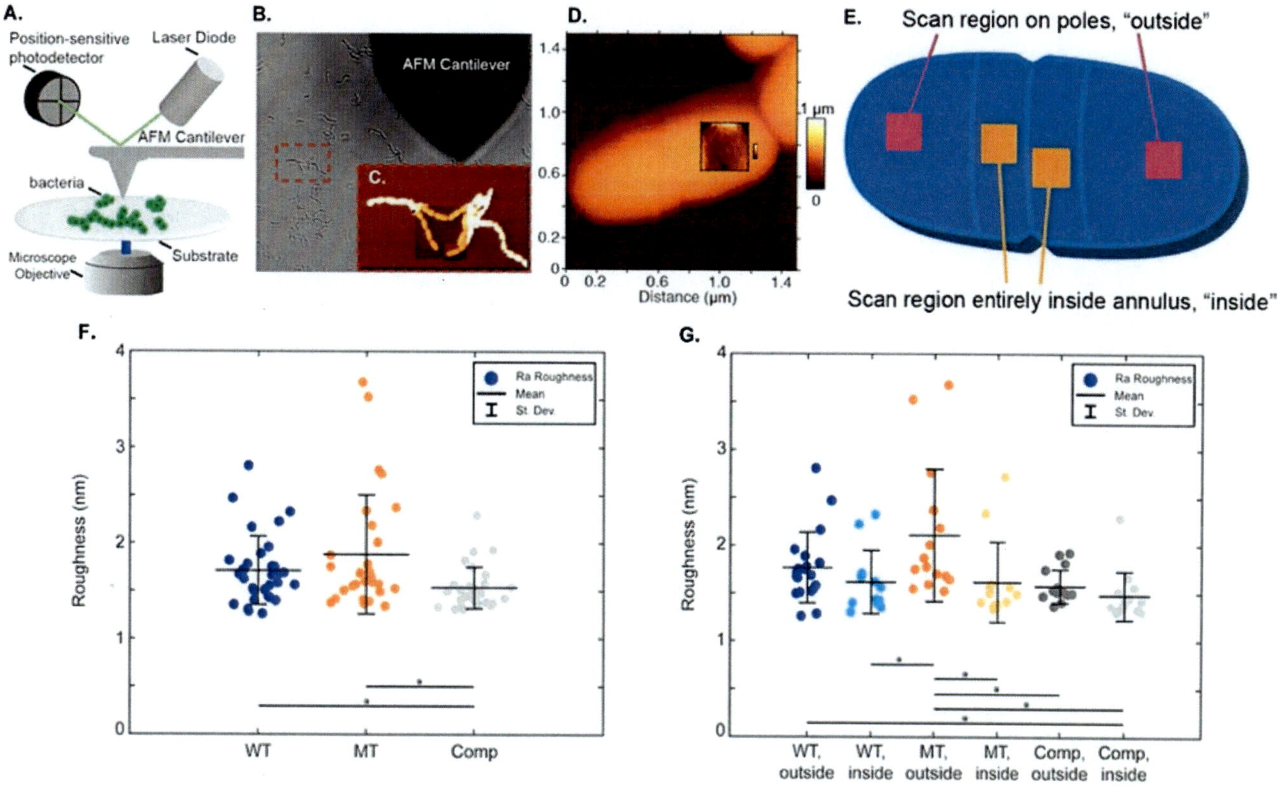

Fig. 15.1 (**a**) Schematic of AFM and confocal instrument setup. (**b**) Brightfield image of bacteria on substrate with identified AFM scan region in (**c**). (**c**) Non-optical AFM image of scanned region shown in (**b**). (**d**) 1.5 μm by 1.5 μm scan to pinpoint single bacterium with 300 nm by 300 nm scanned region utilized for surface roughness determination. (**e**) Schematic depicting "outside" and "inside" bacteria cell scan regions. (**f**) Quantitative surface roughness measurements for WT, MT, and Comp (error bars represent standard deviation). (**g**) Quantitative surface roughness measurements for inside and outside regions for WT, MT, and Comp (error bars represent standard deviation)

By conducting this series of experiments, force curves are collected that allow for the extrapolation of measured height values, and surface roughness. Proper aseptic technique, consistency, and effective sample preparation are necessary to performing successful confocal and atomic force microscopy. Poly-l-lysine is used to immobilize and paraformaldehyde encourages cross-linking of the bacteria on the glass substrate [10–12].

The sample and data collection procedure are highly selective in determining our dependent variable, surface roughness. The comparison between the three strains is made between various locations on single bacteria to avoid biases and to quantify binary fission effects on the cell wall. Binary fission is the method in which bacteria reproduce [13]. The focus of the experiment was to examine the region that lies between a bacterium septum, where binary fission occurs, and the bacterium annulus, where binary fission began. A possible reason for this difference stems from "older" and "newer" cell wall growth, which is dependent on numerous factors, including strain type and temperature [14]. The method in which bacteria reproduce may have significance on the cell wall polymer chain alignment and associated mechanical properties [15]. Figure 15.1(E.) shows an illustration of these regions for better understanding. It is important that the scans do not contain septal regions (found at the center of a bacterium) or annuli (deviated outward from the septum) to avoid skewing the data. The experiment focused on characterizing the smooth cell wall near the poles of the bacterium, which will be referred to as an "outside" scan, and in between the septum and annulus, which will be referred to as an "inside" scan.

From these experiments, 32, 28, and 30 successful scans were achieved for WT, MT, and Comp, respectively, and average surface roughness (Ra) was measured for each (seen in Fig. 15.1(F.)). The number of outside/inside iterations for the three strains, are 20/12 (WT), 13/15 (MT), and 16/14 (Comp). Figure 15.1(G.) contains the Ra for outside and inside scans of each strain. The error bars represent the standard deviation and each dot is one iteration from the successful scans. Student's t-test was applied to determine significant relationships ($p < 0.05$) between the strains and the two regions assessed. Surface roughness values were determined as 1.71 nm, 1.88 nm, and 1.53 nm for WT, MT, and Comp, respectively. For the combined set of roughness values, by statistical analysis, WT and MT showed the same surface roughness and Comp was statistically significant from WT and MT, providing two significant relationships. Assessing the near-pole and within the septal regions provided five statistically significant relationships: WT, outside/Comp, inside, WT, inside/MT outside, MT, outside/MT, inside, MT, outside/Comp, outside, and MT, outside/Comp, inside. These findings support that the microscale and nanoscale cell wall modifications do have an effect on the cell wall roughness for the *S. mutans* strains, and, further, it can be distinguished that based on Ra, the outside regions on the MT bacterium appear to be rougher, on average. Possible explanations may lie in the way the cell forms itself during reproduction. The individual polymer chains containing the glycerol phosphates and wall teichoic acids may need more time to grow and align, which might attribute to a rougher, more mature cell wall structure near the pole as compared to near the septum. The statistically significant relationships are denoted by asterisk and bars on Fig. 15.1 (F.-G.).

The study determined that the molecular-sized glycerol phosphate modification had minimal changes on the quantitative surface roughness between the three chosen strains. Perhaps these findings allude to the fact that single-molecule changes between various strains of bacteria are dependent on more than just the glycerol phosphate. Additionally, more information on the genetic modification can be analyzed in elastic modulus for significant relationships, which are quantities that can be measured from the same instrument setup and sample procedure used in this research. The future of this experiment is to investigate double and triple mutant strains to evaluate if an extra or two extra biological physical changes affects the same properties, and then compare. Elastic modulus will also be investigated to see if the peptidoglycan layer alterations change the stiffness of the selected strains.

Acknowledgements We thank the Department of Molecular and Cellular Biochemistry at UK for use of bacterial culture equipment and materials. Atomic force and confocal microscopy were carried out in the UK Light Microscopy Core user facility. The facility provides access to and expertise in a wide range of advanced imaging instrumentation that include confocal, multiphoton, and super resolution microscopes.

References

1. Lambert, P.: Cellular impermeability and uptake of biocides and antibiotics in Gram-positive bacteria and mycobacteria. J. Appl. Microbiol. **92**, 46S–54S (2002)
2. Hughes, A., Hancock, I., Baddiley, J.: The function of teichoic acids in cation control in bacterial membranes. Biochem. J. **132**(1), 83–93 (1973)
3. Gründling, A., Schneewind, O.: Synthesis of glycerol phosphate lipoteichoic acid in Staphylococcus aureus. Proc. Natl. Acad. Sci. **104**(20), 8478–8483 (2007)
4. Coley, J., et al.: The linkage between teichoic acid and peptidoglycan in bacterial cell walls. FEBS Lett. **88**(1), 1–9 (1978)
5. Edgar, R.J., et al.: Discovery of glycerol phosphate modification on streptococcal rhamnose polysaccharides. Nat. Chem. Biol. **15**(5), 463–471 (2019)
6. Kassies, R., et al.: Combined AFM and confocal fluorescence microscope for applications in bio-nanotechnology. J. Microsc. **217**(1), 109–116 (2005)
7. Schaus, S.S., Henderson, E.R.: Cell viability and probe-cell membrane interactions of XR1 glial cells imaged by atomic force microscopy. Biophys. J. **73**(3), 1205–1214 (1997)
8. McNally, H.A., et al.: Comparative three-dimensional imaging of living neurons with confocal and atomic force microscopy. J. Neurosci. Methods. **142**(2), 177–184 (2005)
9. Grady, M.E., Composto, R.J., Eckmann, D.M.: Cell elasticity with altered cytoskeletal architectures across multiple cell types. J. Mech. Behav. Biomed. Mater. **61**, 197–207 (2016)
10. Chopinet, L., et al.: Imaging living cells surface and quantifying its properties at high resolution using AFM in QI™ mode. Micron. **48**, 26–33 (2013)
11. Vadillo-Rodríguez, V., et al.: Comparison of atomic force microscopy interaction forces between bacteria and silicon nitride substrata for three commonly used immobilization methods. Appl. Environ. Microbiol. **70**(9), 5441–5446 (2004)
12. Greif, D., et al.: High resolution imaging of surface patterns of single bacterial cells. Ultramicroscopy. **110**(10), 1290–1296 (2010)
13. Angert, E.R.: Alternatives to binary fission in bacteria. Nat. Rev. Microbiol. **3**(3), 214–224 (2005)
14. Ratkowsky, D., et al.: Relationship between temperature and growth rate of bacterial cultures. J. Bacteriol. **149**(1), 1–5 (1982)
15. Turner, R.D., et al.: Molecular imaging of glycan chains couples cell-wall polysaccharide architecture to bacterial cell morphology. Nat. Commun. **9**(1), 1263 (2018)

16 Characterization of Reversible Tablet Sliding in Nacre from *Haliotis rufescens* (Red Abalone)

Hayden M. Price, Anqi Zhang, and Barton C. Prorok

Abstract

Nacre or "Mother of Pearl" is a biogenic material that has garnered a lot of attention as its hierarchal architecture of calcite-based materials imparts extraordinary mechanical performance compared to its monolithic form. Nacre is a micro-laminate composite composed of layers of organic matrix and calcium carbonate tablets (aragonite). The tablets are polygonal in shape and approximately 400–500 nm thick. They are organized in irregularly stacked columns that overlap each other, often described as stacks of coins randomly overlapping each other. Each tablet is covered in nano-scale asperities that help lock the tablets together. The Nacre architecture has been well studied the past decade or so to ascertain the mechanistic origins of its extraordinary performance. One of the important aspects of its behavior involves tablets sliding relative to one another. This sliding has been well characterized by various mechanical methods, mostly involving tensile and indentation-based methods. A clear assentation of the developed models suggests that this tablet sliding should be reversible to a degree. However, reversible sliding has only been observed in a few instances and not studied systematically. This work reports on a method that was adept at observing wide-scale reversible tablet sliding. A systematic study this phenomenon was performed enabled by the precise deformation control it offered. The results determined a threshold strain whereby the deformation from tablet sliding became permanent, suggesting the failure of the organic matrix's recovery ability.

Keywords

Reversible sliding, Abalone · Nacre

16.1 Introduction

The abalone nacre has been the subject of many investigations due to its special micro-structure. The abalone shell is composed primarily of calcium carbonate ($CaCO_3$) which has been organized into a complex hierarchical stacked tablet structure (Fig. 16.1) through a self-assembly process known as biomineralization. In this process, nanoscale aragonite tablets are formed layer by layer and interconnected with an organic matrix. The main component of this organic matrix is β-chitin, a silk fibroin-like protein [1].

Many mechanisms have been explored to attempt to develop a relationship between the unique structure and the mechanical properties of abalone nacre, such as a mineral bridges connecting between two aragonite tablets [2], nano-asperites on the aragonite tablet's surface [3], the interaction between the organic matrix and aragonite tablets [4], as well as the existence of a mesolayer [5]. However, all these models conclude that the nacre behaves like a typical elastoplastic material, with its yielding induced by the hierarchical structure and not the base materials. In this paper, we report our observation of the

H. M. Price · A. Zhang · B. C. Prorok (✉)
Department of Materials Engineering, College of Engineering at Auburn University, Auburn University, Auburn, AL, USA
e-mail: prorok@auburn.edu

J. Notbohm et al. (eds.), *Mechanics of Biological Systems and Materials & Micro-and Nanomechanics & Research Applications*, Conference Proceedings of the Society for Experimental Mechanics Series,
https://doi.org/10.1007/978-3-030-59765-8_16

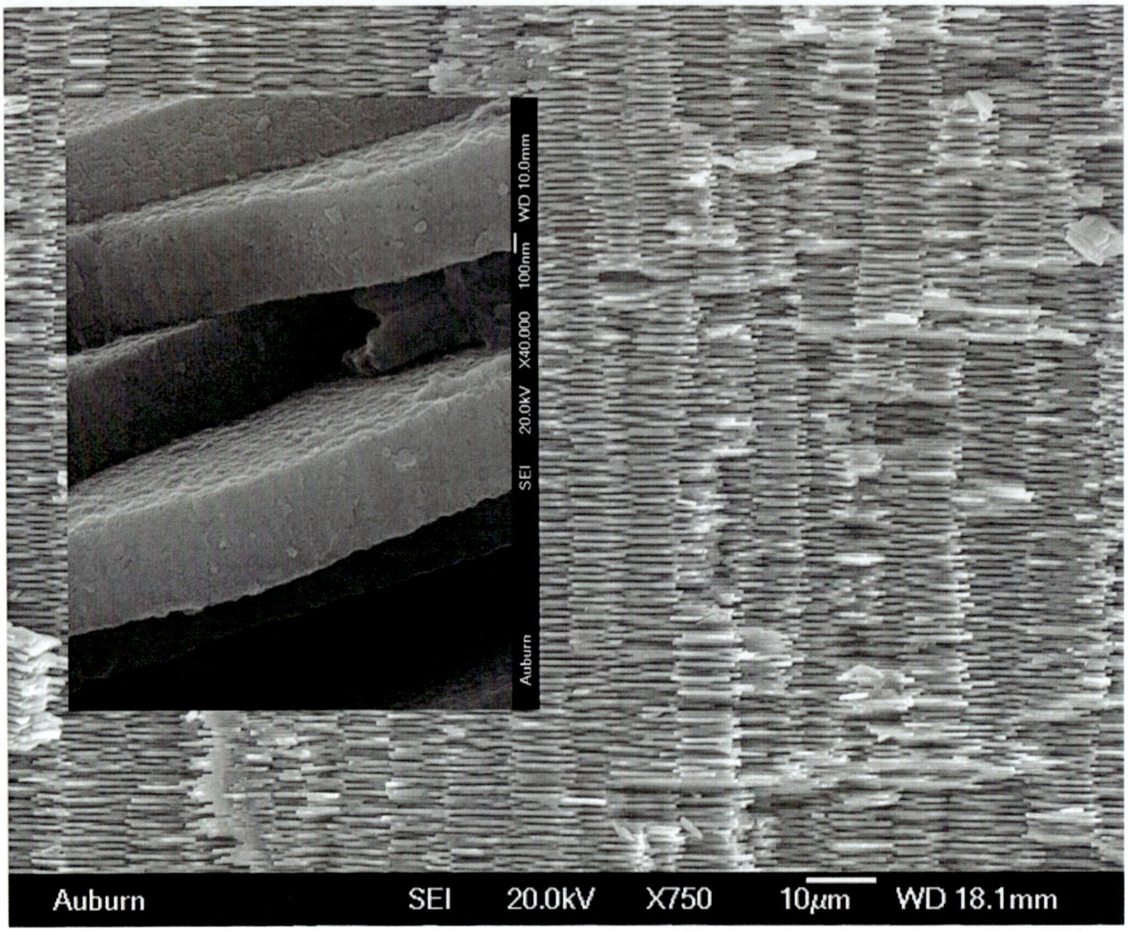

Fig. 16.1 SEM image of the fractured surface of the abalone nacre with a higher magnification image in the upper left corner

Fig. 16.2 Rudimentary testing apparatus for 3-point bend test of nacre beams

reversible sliding of nacre tablets that occurs during the bending and unbending process of the abalone nacre. This finding may be a new source of strengthening mechanism that could help to explain the excellent mechanical properties of the abalone nacre.

16.2 Methods and Materials

Nacre samples were sectioned from the shell of a California Red Abalone. The sections were ground and polished into a thin sheet and then sectioned again into thin beams, with thickness of 400–500 μm and a length of 1.5 cm. These beams then had their cross-sectional surface finely polished so that the tablets motion could be observed under scanning electron microscopy (SEM).

A rudimentary device (Fig. 16.2) was constructed in order to perform a basic three-point bend test on the samples. The device consisted of a stage, two holding pins, and a mechanical screw to apply the bend to the sample. The samples were then subjected to three stages of bending: pre-loading, bending, and release. After each subsequent bending was performed, the device was placed into the SEM and images were taken of the cross-section of the beam.

16.3 Results and Discussion

SEM images were taken of the cross-section of the nacre beam following each subsequent bending stage. Images of the overall deformation at the different bending stages were taken (Fig. 16.3). Images of the tablet structure were also taken at the different bending stages at the same location (Fig. 16.4). At the first bending stage, the image in Fig. 16.4 shows small gaps beginning to appear between neighboring tablets. As the beam was bent further, the gaps between the neighboring tablets increased. The gaps between the tablets disappear upon removal of the bending force.

From the SEM image in Fig. 16.3b, the flexural strain of the bent beam was calculated to be 0.512% ± 0.033%. This value was significantly higher than that measured of nacre under tensile testing (~0.1%) [6]. This order of magnitude difference in strain suggests that after the apparent yield, the nacre can continue to deform in a reversible manner. There are two different explanations for this phenomenon. The first is that individual tablets are shaped like a dovetail, interlocking with one another [7]. This interlocking could provide the driving force needed for the tablets to reversibly slide. The second may be related to the organic matrix which surrounds the nacre tablets. It has a highly ordered, sheet like structure may sustain a large shear deformation without undergoing a significant structural change. Thus, the nacre structure can slide reversibly once the load is released.

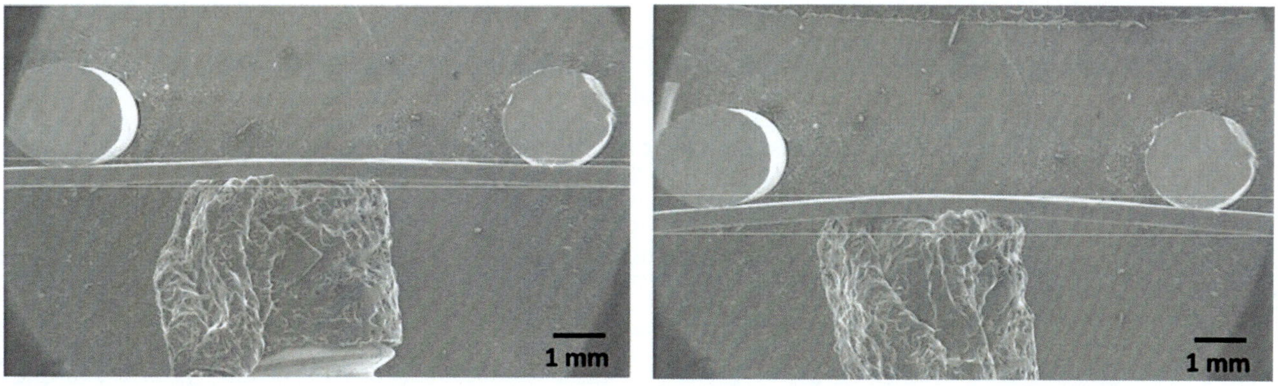

Fig. 16.3 Nacre beam being bent by the device to different degrees, first slightly bent (**a**) then bent even greater (**b**)

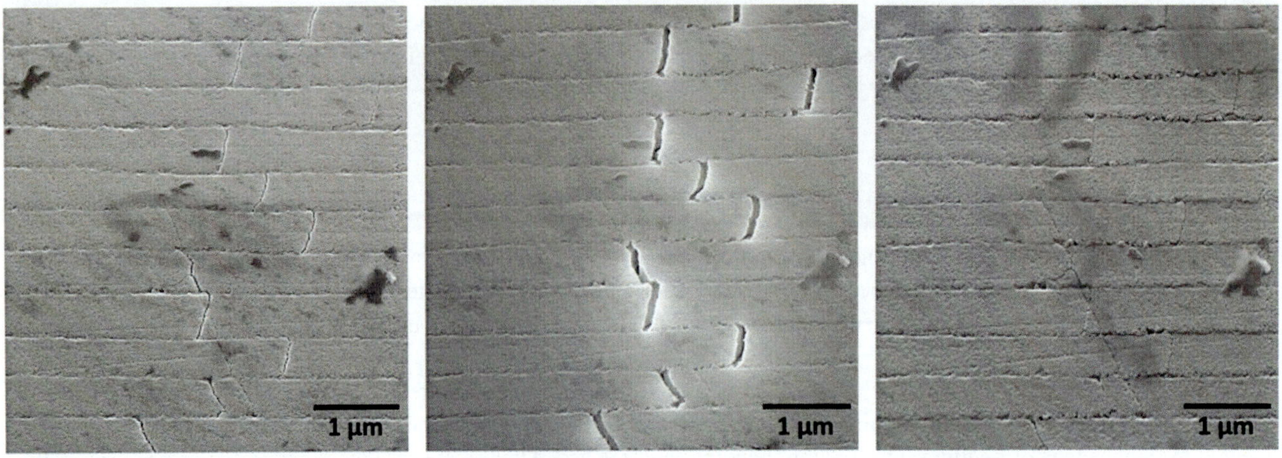

Fig. 16.4 Comparison of tablet separation gap between aragonite tablets of a nacre beam under initial bending (**a**), further bending (**b**), and unloaded (**c**)

16.4 Conclusion

Reversible sliding was observed in the nacre structure of the abalone shell. A permeant relative sliding of nacre tablets was observed in the aforementioned tensile [6] and bending [7] experiment, indicating there is a limited strain that nacre tablets can slide reversibly. This experiment shows the nacre can still deform back to its original form at a lower limit of 0.5% strain. The implications of this observation is that the abalone nacre is more complex than a simple elastoplastic material. The nacre can restore its original structure after undergoing significant straining and even after apparent plastic deformation has occurred. A more precise experiment and theoretical study is needed to establish a better understanding of this new strengthening mechanism. What can be concluded is that while the nacre structure plasticity behaves very differently from that of a typical elastoplastic material, and the stress concentration induced by crack formation in the nacre matrix would be greatly accommodated by this reversible sliding.

Acknowledgements This project was funded through the National Science Foundation (Grant CMMI-1435428).

References

1. Weiner, S., Traub, W., Parker, S.B.: Macromolecules in mollusc shells and their functions in biomineralization. Philos. Transac. R. Soc. Lond. B. **304**(1121), 425–434 (1984)
2. Song, F., Bai, Y.L.: Effects of nanostructures on the fracture strength of the interfaces in nacre. J. Mater. Res. **18**, 1741–1744 (2003)
3. Evans, A.G., et al.: Model for the robust mechanical behavior of nacre. J. Mater. Res. **16**, 2475–2484 (2001)
4. López, M.I., Meyers, M.A.: The organic interlamellar layer in abalone nacre: formation and mechanical response. Mater. Sci. Eng. C. **58**, 7–13 (2016)
5. Sullivan, M., Chen, Y., Prorok, B.: New strengthening mechanisms of nacre in the abalone shell. Int. J. Exp. Comput. Biomech. **3**, 236–249 (2015)
6. Barthelat, F., Tang, H., Zavattieri, P.D., Li, C.-M., Espinosa, H.D.: On the mechanics of mother-of-pearl: a key feature in the material hierarchical structure. J. Mech. Phys. Solids. **55**, 306–337 (2007)
7. Espinosa, H.D., et al.: Tablet-level origin of toughening in abalone shells and translation to synthetic composite materials. Nat. Commun. **2**, 173 (2011)

17 Failure of Three-Tab Shingles Subjected to Wind Gusts Up To 150 MPH: A DIC Based Study

Sreehari Rajan, Troy Myers, and Michael A. Sutton

Abstract

Premature failure of asphalt-shingled roofing systems due to high winds is a major financial issue in the United States. In this study, the mechanism for sealant separation in three-tab shingle systems is evaluated experimentally using (a) in-situ StereoDIC measurements of the deformations in a model shingled roof system subject to wind gusts up to 150 mph and (b) separate fracture experiments for the sealant material to obtain the traction-separation relationship as a function of bonding parameters including temperature and pressure. Results from the series of fracture experiments showed that lower bonding temperatures (55 °C) resulted in partial sealing of the shingles, with separation characterized by adhesive failure of the sealant during wind loading. Shingles with sealant bonding at higher temperature (70 °C) remained intact for more than 2 hours when subjected to wind speeds up to 150 mph.

Keywords

StereoDIC, Asphalt shingle, Sealant adhesive, Wind load, Uplift resistance

17.1 Introduction

Asphalt roof shingles are one of the favorite roofing materials for waterproofing residential and commercial building in North America due to its ease of installation, low maintenance cost and aesthetic appeal. Modern asphalt roof shingles consist of a bitumen-impregnated fiberglass mat with mineral granules on top of the shingles to protect from UV radiation. Figure 17.1 shows a typical modern roof shingle arrangement. As shown in Fig. 17.1, the shingles are secured in place by nailing to the roof panel along the trailing side and by a continuous sealant layer strip near the leading edge of the shingle. The sealant strip material is a form of bitumen, an organic thermoplastic material widely used as a binding material in asphalt pavements and roofing shingles. The sealant layer plays an important role in preventing uplift of the shingle due to wind loads.

Wind resistance ratings are based on experiments, as per ASTM standards such as ASTM D3161 [1] or ASTM D7158 [2]. However, there is a wealth of data today [3, 4] indicating that asphalt roofing systems rated for 150 mph wind speed may incur premature failure under sustained hurricane force winds that are much less than 150 mph. A study by Spark et al. [5] showed that the majority of wind damage to houses during Hurricanes Hugo and Andrew was restricted to the envelop of the building, including roof shingles, windows and doors.

Previous studies by the authors have shown that the shingle and sealant materials are highly viscoelastic at its in-service temperature [6–8]. The fracture toughness of the sealant is also reported to be highly rate dependent, with transition from

S. Rajan (✉) · T. Myers · M. A. Sutton
Department of Mechanical Engineering, University of South Carolina, Columbia, SC, USA
e-mail: sreehari@email.sc.edu

J. Notbohm et al. (eds.), *Mechanics of Biological Systems and Materials & Micro-and Nanomechanics & Research Applications*, Conference Proceedings of the Society for Experimental Mechanics Series,
https://doi.org/10.1007/978-3-030-59765-8_17

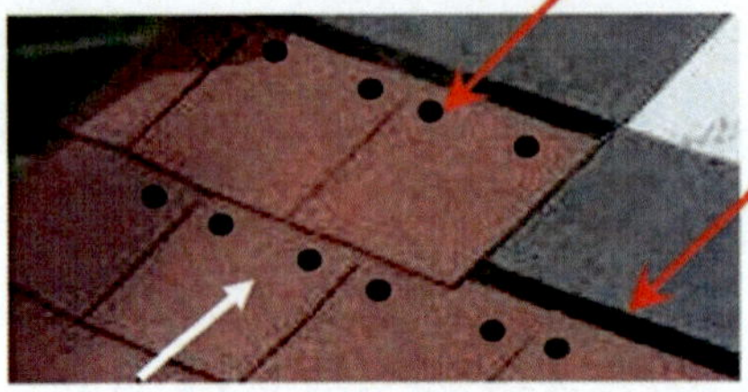

Fig. 17.1 Modern 3-tab asphalt shingles arrangement

ductile to brittle fracture with increasing loading rate [9]. Towards modeling efforts, the authors used a beam on elastic foundation model to show that position of the sealant has an important role in the applied energy release rate [10, 11]. The adhesive in 3-tab shingles is thermally activated. The adhesive achieves its full bond strength if the temperature is above its softening point. Below the softening point, a partial bond is formed. Hence the temperature during bond formation has a significant influence on the performance of 3-tab shingles. In this paper, StereoDIC [12, 13] is used to measure uplift of a two-shingle sealant system formed at two temperatures, one below the softening point of the adhesive and the second one at a temperature above the softening point. Section 2 of the paper includes experimental setup and sample preparation. The results and discussion are presented in Sect. 3. The conclusions are given in Sect. 4.

17.2 Materials and Methods

Commercially available 3-tab shingles randomly drawn from one batch are used in all experiments. The shingle samples are installed using guidelines in the manufacturer's instruction manual. The sealant is activated by placing the shingled plywood test specimen inside an oven for 16 h at two different temperatures, 55 °C and 70 °C in this work. After installing the heat-treated shingled test specimen on a rigid frame[1], the measured uplift displacement of the shingles is solely due to the displacement of the shingles, not frame deformations.

The experimental setup consists of a stereo camera system, tripod for mounting the cameras, LED illumination source and a rigid fixture to mount the shingled plywood test specimen; the pitch angle of the rigid fixture is 10°. The complete experimental setup is shown in Fig. 17.2. The presence of natural high contrast pattern on the top surface of the shingles as shown in the inset to Fig. 17.2 served as a natural speckle pattern for StereoDIC analysis. Correlated solutions VIC 3D[2] software is used for processing speckle images. Table 17.1 shows the StereoDIC parameters used in all image analyses performed.

17.3 Results and Discussion

The shingle-sealant system is subjected to a specific combination of wind speeds

- 60 mph for 600 s
- Ramp up to 90 mph for the next 600 s.
- If the shingles sustained the wind load after the first two stages, wind speed is increased to 150 mph and maintained for 6800 s or till failure of the shingle.

The uplift displacement field for the shingle is continuously measured at a frequency of 1 Hz. The uplift displacement map of two shingles with sealant bonding temperatures (T_b) of 55 °C and 70 °C after 1350 s are shown in Fig. 17.3. From Fig. 17.3 it can be observed that the first shingle sample with a sealant bond formation temperature of 70 °C sustained much

[1]Independent StereoDIC measurements of the deflections of the plywood sheet bolted to the rigid pedestal frame were performed by the authors for wind loading above 150 mph. Results showed that the plywood surface motion was less than 20 μm at maximum wind speed.

[2]https://www.correlatedsolutions.com/vic-3d/

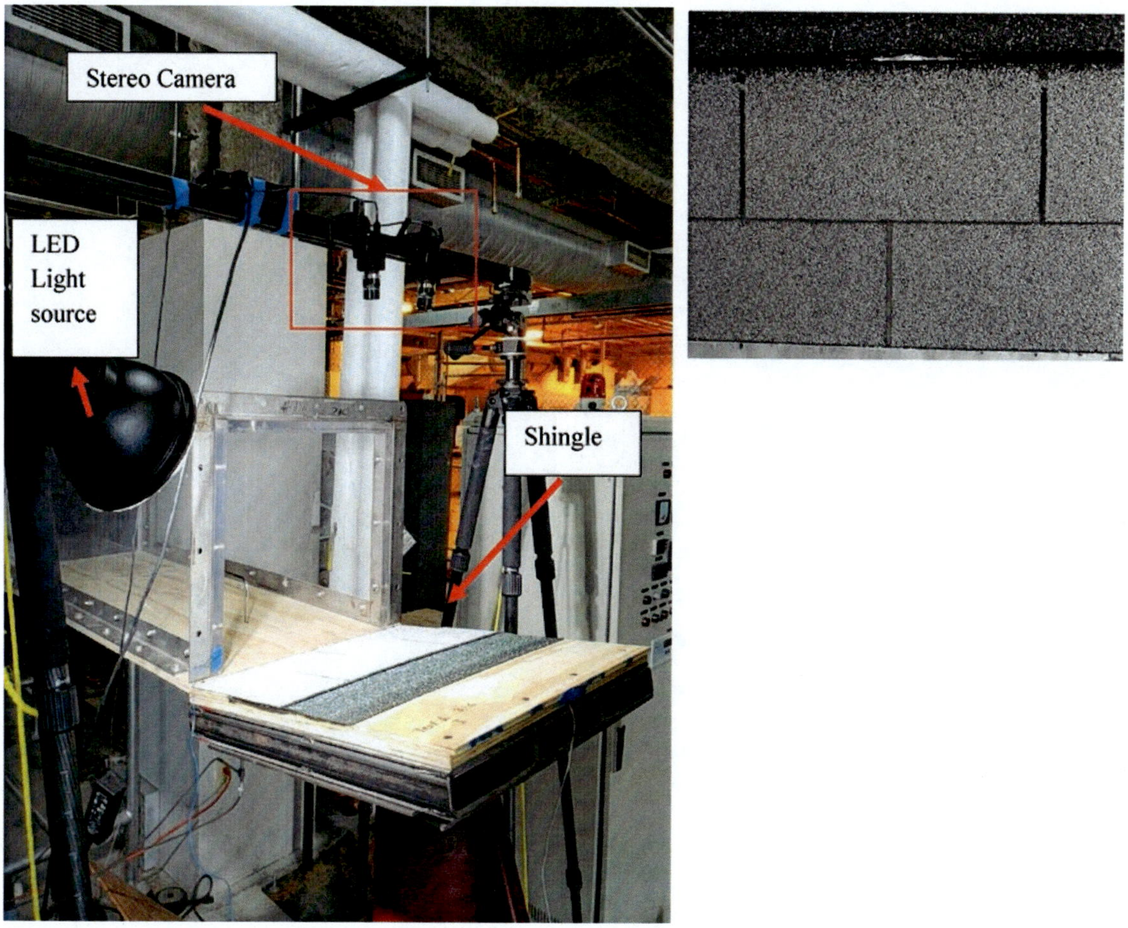

Fig. 17.2 StereoDIC experimental setup; Inset: natural speckle pattern on the top surface of the shingle

Table 17.1 DIC and camera parameters

Subset size	45 × 45 pixels^2
Step size	5 pixels
Filter type	Center-weighted Gaussian filter
Strain filter size	5 × 5 data points (area of 5 × 5 pixels^2)
Camera	5 MP CMOS PointGrey
Lens	35 mm Schneider
Field of view	500 × 500 mm
Magnification	190 μm/ pixel
Average speckle size	1.5 mm

lower uplift along the leading edge compared to the second shingle sample with sealant bond formation temperature of 55 °C. The second shingle sample failed after 1350 s whereas the first sample incurred no damage after 3000 s.

The shingle with sealant bonding at a temperature of 55 °C was not able withstand wind load of 150 mph for more than 120 s, whereas the shingle with T_b = 70 °C sustained no apparent damage and no separation, even after 6800 s of 150 mph speed wind load. Inspection of the failed shingles confirmed that the lower bond formation temperature resulted in partial bonding of the sealant, with failure predominantly in the form of adhesive separation. For the fully bonded shingles ($T_b \geq 70$ °C) the sealant never failed due to the wind load and a peel test on assessing the strength of the bond strength revealed that the failure occurs in the interface between the glass fabric layer and asphalt layer within the top shingle, and not in the sealant layer.

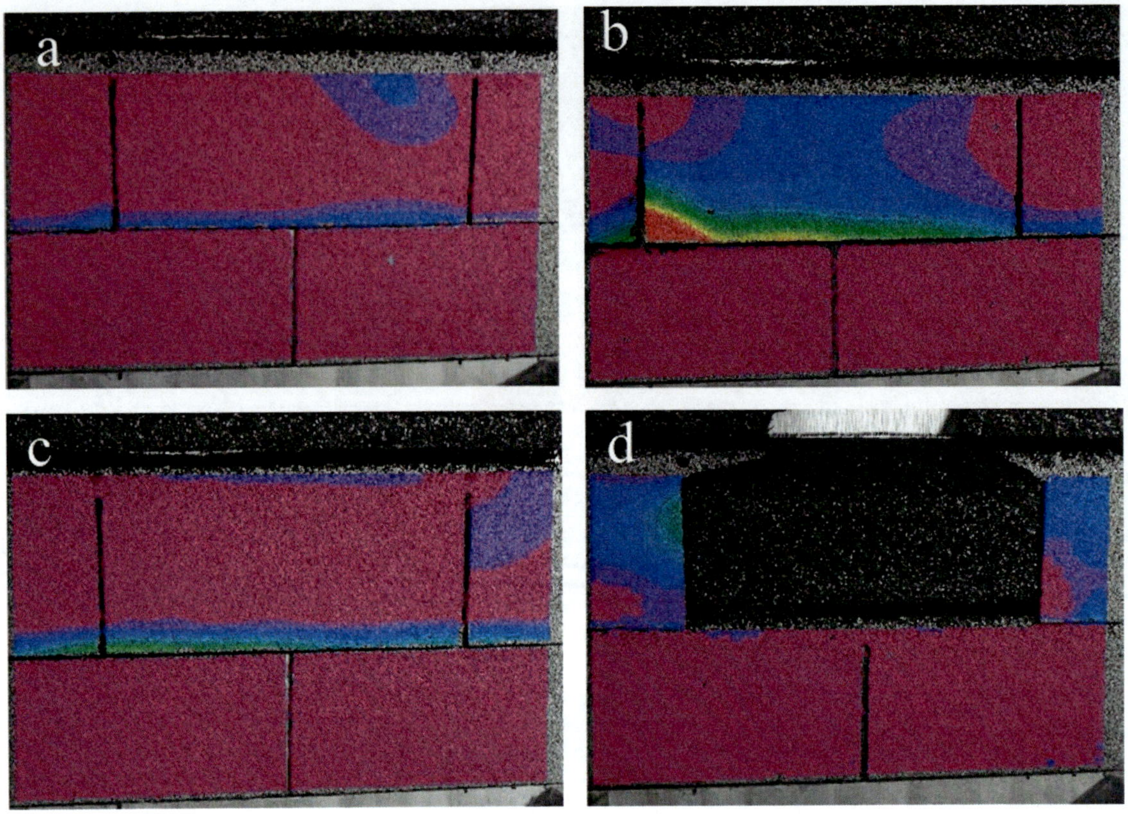

Fig. 17.3 Uplift displacement map of shingle with (**a**) $T_b = 70$ °C and t = 1320 s, (**b**) $T_b = 55$ °C and t = 1320, (**c**) $T_b = 70$ °C and t = 3000 s, (**d**) $T_b = 55$ °C and t = 3000 s

17.4 Conclusions

Full field measurement of the uplift of 3-tab shingles subjected to wind speed up to 150 mph is performed using StereoDIC. The sealant formed at a temperature below the softening point of the sealant (55 °C) resulted in partial bond formation. For partially sealed adhesives formed at lower temperature, the sealant is the weakest link in the system. Adhesive failure is observed to occur between the sealant strip and top shingle during wind loading. When the sealant is formed at higher temperature, the sealant achieved its full bond strength. For fully bonded shingles, the weakest link is the interface between glass fabric layer and asphalt layer within the top shingle. The partially sealed adhesive sustained a wind speed of 150 mph for 120 s and separated, whereas the fully bonded sealant formed at 70 °C sustained 150 mph wind for more than 6800 s without separation.

Acknowledgements The support of the Department of Mechanical Engineering and the State of South Carolina Center for Mechanics, Materials and NDE, including financial support for graduate and undergraduate students, open access to the wind tunnel and machining/fabrication of the nearly rigid support structure for the shingle system, are gratefully acknowledged.

References

1. Standard test method for wind-resistance of asphalt shingles (fan-induced method), ASTM D3161 American Society for Testing and Materials International, West Conshohocken, PA (2009)
2. Standard test method for wind resistance of asphalt shingles (uplift force/uplift resistance method), ASTM D7158. American Society for Testing and Materials International, West Conshohocken, PA (2011)
3. Charley, H.: Nature's force vs. structural strength. Institute for Business & Home Safety, Tampa, FL (2004)
4. Liu, Z.Z., Pogorzelski, H., Masters, F., Tezak, S., Reinhold, T.: Surviving nature's fury: performance of asphalt shingle roofs in the real world. Interface, 29–44 (2010)

5. Sparks, P.R., Schiff, S.D., Reinhold, T.A.: Wind damage to envelopes of houses and consequent insurance losses. J. Wind Eng. Ind. Aerodyn. **53**(1–2), 145–155 (1994)
6. Rajan, S., Sutton, M.A., Oseli, A., Emri, I., Matta, F.: Linear viscoelastic creep compliance and retardation spectra of bitumen impregnated fiberglass mat and polymer modified bitumen. Constr. Build. Mater. **155**, 664–679 (2017)
7. Rajan, S., Sutton, M.A., Fuerte, R., Emri, I.: Time temperature superposition and prony series coefficients of asphalt roof shingle material from viscoelastic creep testing. In: Arzoumanidis, A., Silberstein, M., Amirkhizi, A. (eds.) Challenges in mechanics of time dependent materials Conference Proceedings of the Society for Experimental Mechanics Series, vol. 2, pp. 33–38. Springer, Cham (2018)
8. Ghorbani, R., Zhao, X., Matta, F., Sutton, M.A., Kidane, A., Liu, Z., et al.: Feasibility of non-contacting measurement of wind-induced full-field displacements on asphalt shingles, pp. 415–421. Springer, Cham (2015)
9. Rajan, S., Sutton, M.A., Fuerte, R., Kidane, A.: Traction-separation relationship for polymer-modified bitumen under Mode I loading: double cantilever beam experiment with stereo digital image correlation. Eng. Frac. Mech. **187**, 404–421 (2018)
10. Croom, B.P., Sutton, M.A., Zhao, X., Matta, F., Ghorbani, R.: Modeling of asphalt roof shingle-sealant structures for prediction of local delamination under high wind loads. Eng. Struct. **96**, 100–110 (2015)
11. Croom, B.P., Rajan, S., Matta, F., Aleshin, A., Myers, T., Sutton, M.A.: Modeling of asphalt roof shingle structures with dual sealant strips; optimization for improved delamination resistance under high wind loads. J. Build. Eng. **30**, 101266 (2020)
12. Sutton, M.A., McNeill, S.R., Helm, J.D., Chao, Y.J.: Advances in two-dimensional and three-dimensional computer vision Photomechanics, pp. 323–372. Springer, Berlin, Heidelberg (2000)
13. Sutton, M.A., Matta, F., Rizos, D., et al.: Recent progress in digital image correlation: background and developments since the 2013 W M Murray Lecture. Exp. Mech. **57**, 1–30 (2017)

Check for updates

Design and Rapid Prototyping of Fiber-optic Based Micro-force Sensors by Two-photon Polymerization

18

Daniel Ruiz-Cadalso, Jeffrey T. Cheng, John J. Rosowski, and Cosme Furlong

Abstract

We are developing fiber-optic based micro-sensors for measurement of dynamically induced micro- to nano-Newton forces. The force-sensing transduction mechanism is based on a series of mechanical micro-springs and an optically reflective micro-plate with axial movement that are embedded at the distal end of single-mode optical fibers. Micro-sensors have low-stiffness, which is required for characterization of biomedical samples and soft materials, and are rapid prototyped by two-photon polymerization methods. Total-internal-reflection based micro-retroreflectors are designed to direct the fiber output laser beam back into its core for subsequent quantification of force-displacement induced phase changes that are measured by interferometric methods. After conducting investigations to characterize mechanical properties of solidified photoresists as a function of fabrication parameters, that include laser exposure, we are designing and tuning the springs to perform over specific ranges of forces and displacements. The mechanical springs are characterized and calibrated by MEMS micro-force sensing probes attached to motorized nano-positioners to detect forces from 100 nano-Newtons to 500 micro-Newtons. We report the performance of the newly developed micro-force sensors as applied to several applications, including measurements of acoustically induced forces on the Tympanic Membrane surface to better understand hearing and middle-ear mechanics.

Keywords

Micro-sensors, Fiber-optics, Rapid prototyping, Two-photon polymerization, Hearing mechanics

D. Ruiz-Cadalso (✉)
Center for Holographic Studies and Laser micro-mechaTronics—CHSLT, Worcester, MA, USA

Mechanical Engineering Department, Worcester Polytechnic Institute, Worcester, MA, USA
e-mail: daruizcadalso@wpi.edu

J. T. Cheng · J. J. Rosowski
Eaton-Peabody Laboratory, Massachusetts Eye and Ear, Boston, MA, USA

Department of Otolaryngology, Harvard Medical School, Boston, MA, USA

C. Furlong
Center for Holographic Studies and Laser micro-mechaTronics—CHSLT, Worcester, MA, USA

Mechanical Engineering Department, Worcester Polytechnic Institute, Worcester, MA, USA

Eaton-Peabody Laboratory, Massachusetts Eye and Ear, Boston, MA, USA

Department of Otolaryngology, Harvard Medical School, Boston, MA, USA

J. Notbohm et al. (eds.), *Mechanics of Biological Systems and Materials & Micro-and Nanomechanics & Research Applications*, Conference Proceedings of the Society for Experimental Mechanics Series,
https://doi.org/10.1007/978-3-030-59765-8_18

18.1 Introduction

Force-sensing technologies are widely used in most engineering industries and applications. From measuring the human body weight to determining the wave characteristics of nanoscale vibrations, these technologies sense the forces that occur from interactions of objects with one another. There is a wide variety of force-sensing methods, and their advantages and disadvantages depend mostly on the measurement scale. Microscale physical forces are popularly sensed with micro-electromechanical systems (MEMS) sensors, which use a combination of springs, cantilever and flexural beams with either strain gauges or thermal displacement sensors. MEMS sensors are among the most popular in engineering industries, used for purposes such as nanoindentation, tensiometers, micro-handling and assembly, etc.

Other popular microscale force sensors include both the fiber Bragg grating and the Fabry-Perot interferometer, which use the indirect interaction between photons and external forces. This method was recently used to develop a fiber-optic based force sensor via direct laser writing [1], with sensing characterization and calibration done by spectroscopic means. Here, we provide a solution to a fiber-optic based microscale force sensor with high repeatability and low fabrication time by rapid prototyping using a two-photon polymerization tool. Two-photon polymerization is type of direct laser writing (DLW) based on non-linear two-photon absorption process. The process enables to fabricate polymer three-dimensional micro- and nano-structures in photosensitive materials. The tool is very popular in research areas within applications like photonics, biomedical engineering, microfluidics, micromechanics, etc. Similar work has proven the feasibility of the fabrication method, with the development of a miniaturized optical fiber microphone [2] and a monolithic force-sensitive 3D microgripper [3].

Fiber-optic based force sensors are suitable especially for soft materials and in-situ sensing applications. With a low-stiffness sensor that can provide a relatively wide range of displacements, the forces seen in mechanical waves travelling on the surface of a skin tissue upon impact-indentation of pressurized air can help to better understand the tissue mechanics. Micro-scale forces inside of the human body can potentially be sensed during MRI scan procedures, since these polymer-based sensors are advantageous against the metallic components which cannot be used in an MRI. Finally, understanding the rupture mechanics of the human tympanic membrane by adding collected vibration force data to the already-studied displacement fields done by high-speed holographic methods [4].

18.2 Mechanical Properties of Solidified Photoresists

Since two-photon polymerization is a technique used in relatively new technologies, there is a lack of substance in research regarding the mechanical properties of photoresists when solidified after exposure to laser. Thus, to fully understand the obtainable mechanical properties of polymerized components, their relationship with the laser exposure settings must be researched and understood. For the purpose of this project, the mechanical properties will only be characterized for the exposure parameters that we can use to successfully fabricate the desired geometrical features.

The experimental tests we designed will use force-induced compressions of uniquely designed testing artifacts to plot force-displacement data to understand the elastic modulus, as seen in Fig. 18.1. The compressions were applied with the use of a MEMS microforce probe attached to a nano-positioner for controlled displacements. Figure 18.1a shows the testing artifacts, which were designed to provide both a stable test during loading and unloading, and more importantly linear deflections for simplified analysis. The design consists of three main sections: a base, deflecting beams, and a loading axle. The two "base blocks" on the lateral extremes provide a stable fixture during loading and unloading occasions, and whose own cantilever-like deflections are minimal and negligible in comparison to the deflecting beams. The controllable input factors of this experimental design will be the thickness of these deflecting beams. Lastly, the central vertical block is the loading axle, where the top surface is the area in which the external force-induced compression makes contact. Figure 18.1b shows the force-displacement results for one of the testing artifacts. The determined elastic modulus of the material for the used fabrication exposure parameters is 2.45 ± 0.1 GPa. The uncertainty is calculated with the use of the error propagation formula, which considers the analytical expression of the plate deflections.

18.3 Design and Fabrication

The force-sensitive mechanism is designed to function on the tip of an optical fiber. It is designed to be sensitive to external forces, while interacting with laser light that is exiting the fiber core. We use single-mode optical fibers that operate within the visible range of wavelengths. The system we designed is an assembly with three main components: a spring loading system, a top plate, and a reflecting mechanism. A schematic of the design and setup can be seen in Fig. 18.2. The design in

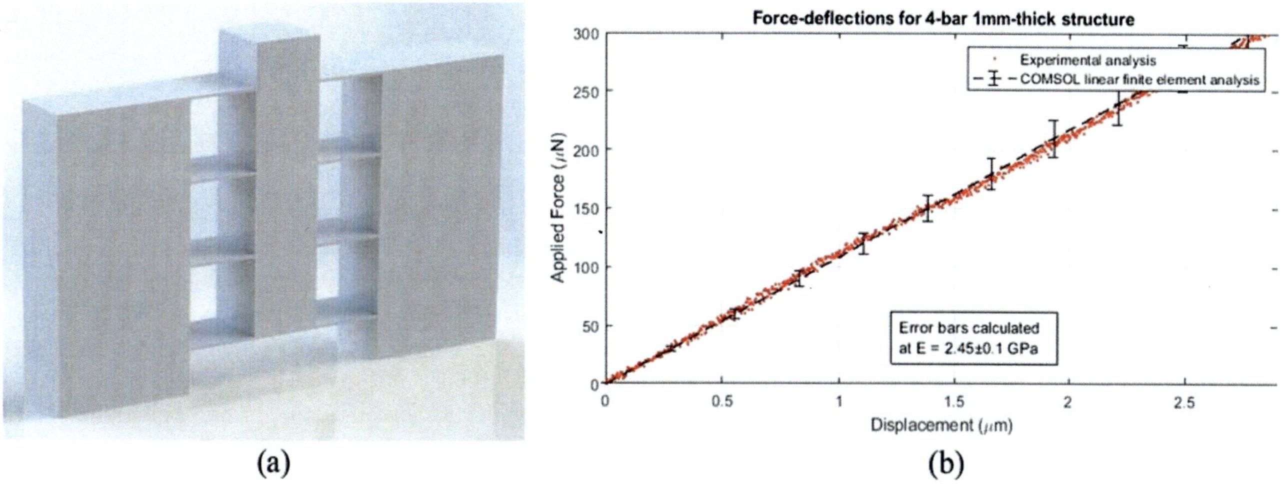

Fig. 18.1 Experimental tests for understanding the Elastic Modulus of the solidified photoresists from fabrication procedures with; (**a**) the designed testing artifact, and (**b**) force-displacement results for induced compressions

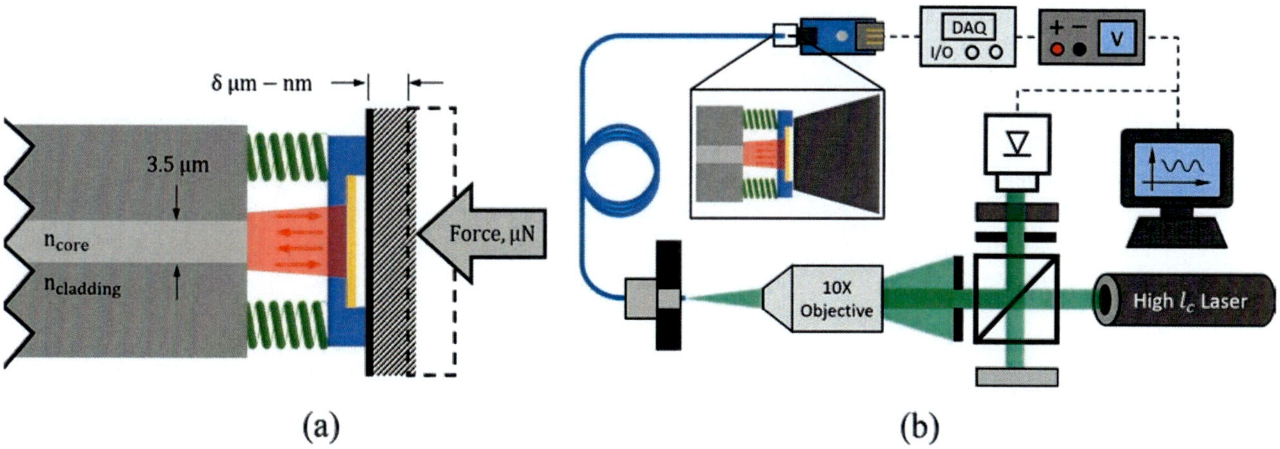

Fig. 18.2 Schematic diagrams of; (**a**) the designed distal-end force-sensing mechanism, and (**b**) the interferometric experimental setup for analysis of force sensor

Fig. 18.2a shows the spring loading system in orange, the top plate in blue, and the reflectors in green, while the shaded area in red represents the light propagating from the end face of the single mode fiber. The light reflects and couples back into the fiber for subsequent analysis at a detector. When an axial force is exerted onto the top plate, the springs will be compressed, which will decrease the optical path length of the reflected light. As a result, the deflection causes a phase shift in the interference pattern between the reflected light and the reference beam at the detector.

Plate deflections due to external induced forces are analyzed through interferometric methods. We constructed a fiber-optic interferometry setup, seen in Fig. 18.2b, that uses a 50:50 non-polarizing beam splitter to divide a beam generated by a laser into two, one being the incidental beam that interacts with the designed mechanism while the other is a reference beam. The reference beam is then reflected by a mirror and is directed straight to the photodetector. The incidental beam is coupled into the optical fiber with the use of a 10X objective and a precision fiber aligner. It then gets transmitted to the mechanism, interacting with the reflecting system. Once reflected, it travels back to the beam splitter, which re-directs it to the photodetector, interfering with the reference beam.

Upon interfering with the reference beam at a detector, a sinusoidal intensity pattern is generated due to temporal coherence. This phenomenon occurs upon the correlation of a wave with itself at different instances in time. If this interval of time difference between both beams is denoted by τ, while the time the light takes to travel its own coherence length l_0 is τ_0, then the interference profile can be described by,

$$I(\tau) = A_0^2 + B_0^2 + 2A_0B_0\Lambda(\tau/\tau_0)\cos(\Delta\phi), \qquad (18.1)$$

where A_0 and B_0 are the amplitudes of the superimposed optical fields, and $\Lambda(\tau/\tau_0)$ describes the contrast.

18.4 Results and Discussions

Two-photon polymerization (2PP) 3D micro productions of the components were accomplished using the Nanoscribe Photonic Professional GT+ (Nanoscribe GmbH, X). The configuration used includes a dip-in approach with the 63X microscope objective in IP-Dip acrylate 2PP resin with a 0.7 mm-thick fused silica (FS) substrate. The scan mode in the X-Y uses the integrated mirror galvanometers in the system, while the Z-position is auto-adjusted with the system's piezo. The laser power for the system is calibrated at 50 mW. Using a commercial bare fiber holder integrated with a custom-made mount, we fabricate the mechanism onto the tip of a single mode optical fiber. Figure 18.3 shows the experimental calibration of the force-sensing system. The mechanism is compressed and uncompressed with a commercial MEMS micro-force sensing probe, as seen in Fig. 18.3a. The sensor probe is attached to a nano-positioner and controlled to compress the functionalized fiber at a rate of 2 μm/s. The voltage data from the force sensor is acquired, analyzed and converted to force data. We then convert the time axis to displacement using the rate of displacement and acquire the force-displacement plot of the fabricated force-sensing mechanism, as shown in Fig. 18.3b.

Prior to the test and characterization of the complete system, we conducted proof-of-concept experiments. The force-sensing system was fabricated without the reflecting mechanism and tested with force-induced compressions using a mirror attached to a nano-positioner. Upcoming tests, however, will see a piezo replacing the nano-positioner for higher control and precision of displacements. Figure 18.4 shows fabrication and experimental results of the force-sensing mechanism. We fabricated the force-sensing system at the tip of the optical fiber with the addition of the reflection mechanism for representation of results. A Scanning Electron Microscope (SEM) image of the functionalized fiber is seen in Fig. 18.4a. Another mechanism was fabricated without the retroreflectors to obtain proof-of-concept results using a mirror, which are seen in Fig. 18.4b. During the test, the mirror compressed the mechanism at a rate of 0.5 μm/s, and the intensity data at the photodetector was taken and processed.

Acknowledgements Thanks to our collaborator Jeffrey T. Cheng from Eaton-Peabody Laboratory, Massachusetts Eye and Ear, as well as technical support from CHSLT graduate students who aided in completing this project. Partial support from the NIH-NIDCD and the DoE are also acknowledged.

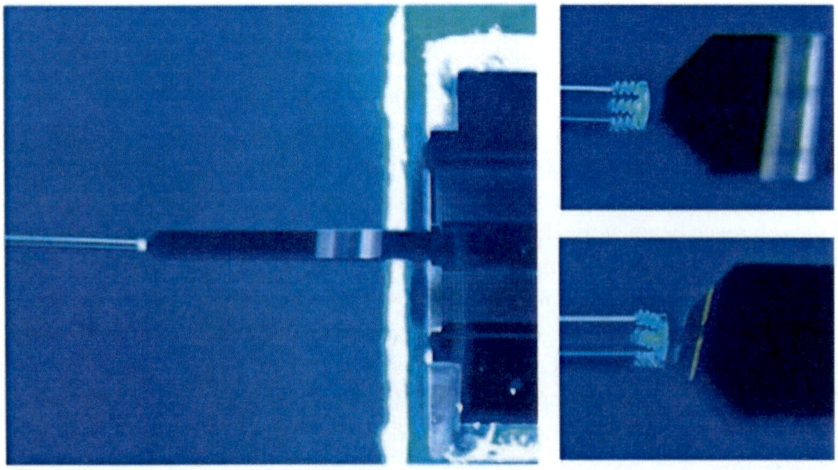

(a)

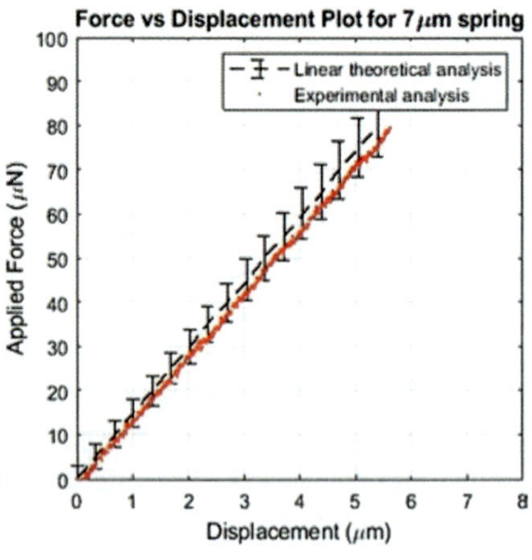

(b)

Fig. 18.3 Experimental calibration of force-sensing mechanism, consisting of; (**a**) microscopic images of compressed and uncompressed states, and (**b**) computational characterization of force-displacement data

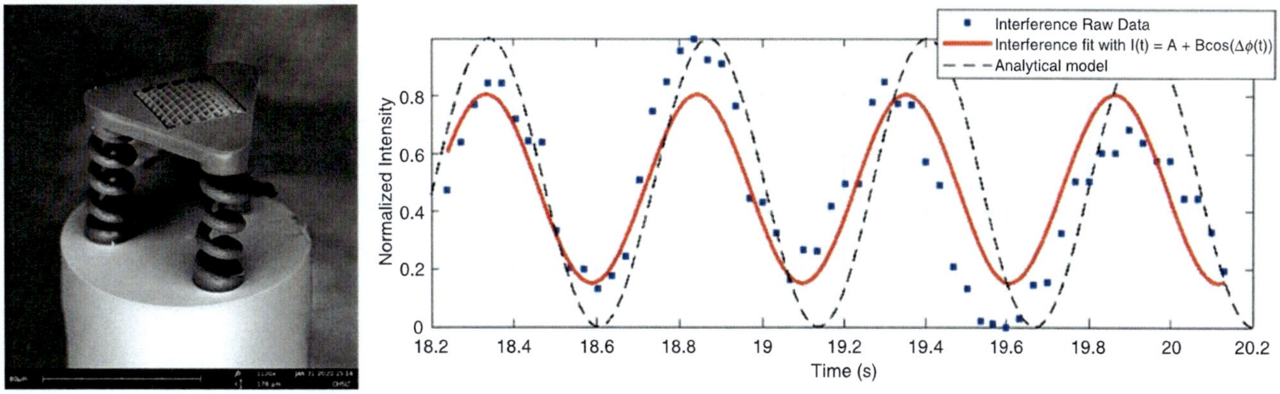

(a) (b)

Fig. 18.4 Fabrication and experimental results of; (**a**) force-sensing of proof-of-concept results of force-induced compression of mechanism using a mirror

References

1. Thompson, A.J., Power, M., Yang, G.Z.: Micro-scale fiber-optic force sensor fabricated using direct laser writing and calibrated using machine learning. Opt. Express. **26**(11), 14186–14200 (2018)
2. Wang, H., Xie, Z., Zhang, M., Cui, H., He, J., Feng, S., Wang, X., Sun, W., Ye, J., Han, P., Zhang, Y.: A miniaturized optical fiber microphone with concentric nanorings grating and microsprings structured diaphragm. Opt. Laser Technol. **78**, 110–115 (2016)
3. Power, M., Thompson, A.J., Anastasova, S., Yang, G.Z.: A monolithic force-sensitive 3D microgripper fabricated on the tip of an optical fiber using 2-photon polymerization. Small. **14**(16), 1703964 (2018)
4. Tang, H., Razavi, P., Pooladvand, K., Psota, P., Maftoon, N., Rosowski, J.J., Furlong, C., Cheng, J.T.: High-speed holographic shape and full-field displacement measurements of the tympanic membrane in normal and experimentally simulated pathological ears. Appl. Sci. **9**(14), 2809 (2019)

Experimental Study of Shear and Tensile Properties of LIGA Ni-Fe and Ni-Co Alloys at Quasi-static and Intermediate Strain Rates

19

Li-Anne Liew, David T. Read, May L. Martin,
Todd R. Christenson, Peter E. Bradley, Nicholas Barbosa,
Frank W. DelRio, Jeffrey R. Smyth, and John T. Geaney

Abstract

We report the in-plane shear properties, to fracture, of 200 μm-thick LIGA nanocrystalline Ni-10%Fe and micrograined Ni-10%Co. Planar simple shear specimens were designed, fabricated and pulled to fracture using a miniature commercial tension test stage. Forces were measured with a load cell and displacements were measured via digital image correlation. A procedure was developed to extract the shear strains from the displacements. We show shear stress vs shear strain curves to fracture for each material, at strain rates 0.001/s and 30/s. Tensile tests were also performed on both materials at strain rates 0.001/s and 1/s. Increases in shear strength, tensile strength, and true strain to fracture were observed at the higher strain rate, but rate effects were inconclusive for the Young's modulus, shear modulus, and yield strength due to large measurement uncertainties. Finally, the stress triaxiality and equivalent plastic strain at fracture for the tensile and shear specimens were estimated and used to construct part of the fracture locus for the Ni-10%Fe material. It is anticipated that this data may be used to facilitate failure prediction models for LIGA MEMS.

Keywords

Electrodeposited nickel, LIGA · MEMS, Shear, Fracture locus

19.1 Introduction

The commercialization of Micro Electro Mechanical Systems (MEMS) in a growing number of industries has resulted in an increased emphasis on MEMS structural reliability and accurate failure prediction models. To this end, material data under various loading conditions and with well-understood uncertainties are needed to construct and validate material constitutive models. Many works over the past two decades have reported the mechanical properties of films of Ni and its alloys fabricated through the LIGA process. LIGA [1, 2] is a German acronym for Lithographie, Galvanoformung, Abformung, meaning lithography, electroplating and molding. However, while tensile, fatigue, bending, creep and fracture toughness properties

L.-A. Liew (✉) · D. T. Read · M. L. Martin · P. E. Bradley · N. Barbosa
Applied Chemicals and Materials Division, National Institute of Standards and Technology, Boulder, CO, USA
e-mail: li-anne.liew@nist.gov

F. W. DelRio
Sandia National Laboratories, Albuquerque, NM, USA

T. R. Christenson
HT MicroAnalytical, Albuquerque, NM, USA

J. R. Smyth · J. T. Geaney
U.S. Army Combat Capabilities Development Command Armaments Center, Picatinny Arsenal, NJ, USA

J. Notbohm et al. (eds.), *Mechanics of Biological Systems and Materials & Micro-and Nanomechanics & Research Applications*, Conference Proceedings of the Society for Experimental Mechanics Series,
https://doi.org/10.1007/978-3-030-59765-8_19

of LIGA alloys have been given attention in the literature, far fewer papers to date report their shear properties to fracture or the fracture locus of LIGA materials. Two examples of shear property measurements include Refs [3, 4] which reported the shear modulus of silicon and LIGA Ni using torsional specimens that were twisted out of plane by bench-top loading apparatuses, and the angular displacement was measured optically, but while the torsional test of the silicon specimens resulted in fracture, the test on the LIGA Ni was not to fracture. In-plane torsional tests have been performed, to fracture, on single crystal silicon pillars with cross sectional areas of about 1 um^2, through use of in-situ fabricated silicon MEMS actuators for applying the loads [5]. Micro-torsional tests have been used to measure the shear modulus of other MEMS materials, see for example the recent review of micro-mechanical test techniques in Ref [6]. With micro-torsional test techniques, however, shear tests, especially to fracture, are technically challenging and typically not performed for ductile materials, and shear strains are difficult to measure when there is extensive plastic deformation.

We conducted a detailed experimental study of the in-plane shear properties, to fracture, of two LIGA alloys: a high-strength nanocrystalline Ni-10%Fe alloy and a ductile micrograined Ni-10%Co alloy. Unlike the previous work on the LIGA Ni shear modulus and most of the MEMS materials literature, our shear test approach did not involve torsional loading. Instead, we drew from the tradition of macroscale sheet metal studies and used planar specimens (200 μm thick, and with gauge width of 600 μm) that were loaded in tension to produce near-pure shear stresses in the center of the specimen. Since the LIGA fabrication process is planar and results in anisotropy in microstructure and properties due to electrodeposition, LIGA MEMS devices under complex loading are likely to experience in-plane shear stresses, thus planar specimens would more accurately capture the material behavior in these conditions. Other advantages of planar specimens over micro-torsional specimens are that the stress state stays constant to larger deformations, tests to fracture and at various strain rates can be more easily performed, and shear strength and yield strength in shear can be directly measured. We also conducted tensile tests on specimens with gauge sections 2 mm long × 200 μm wide × 200 μm thick, with more details in Refs. [7, 8]. We report here the shear and tensile properties of the two alloys, at two strain rates, and show part of the fracture locus of the nanocrystalline Ni-10%Fe alloy, constructed from the fracture strains of the tensile and shear specimens. This data can potentially be used to construct or validate failure prediction models for LIGA MEMS.

19.2 Experimental Methods

The textbook definition of simple shear for a material of width, W, along the y-axis, is illustrated in Fig. 19.1a for the xy Cartesian coordinate system. When a displacement, $u = \Delta x$, is applied, an angle, α, is formed. The simple shear strain, γ, is given by:

$$\gamma = \tan(\alpha) = \frac{\Delta x}{W} \tag{19.1}$$

Much work has been done over the past several decades in developing and analyzing designs of planar simple shear specimens of various materials and developing test procedures at the macroscale. We based our specimen design on a scaled down version of ASTM B831 [9], which is a standard for shear testing of sheet metals. However, the ASTM design, shown in Figs. 19.1b, c, calls for the gauge section to be thinner than the rest of the specimen. The width of the reduced section ("W" in Fig. 19.1c) can be taken as W in Fig. 19.1a and Eq. 19.1. Our specimens, on the other hand, were designed as a single layer

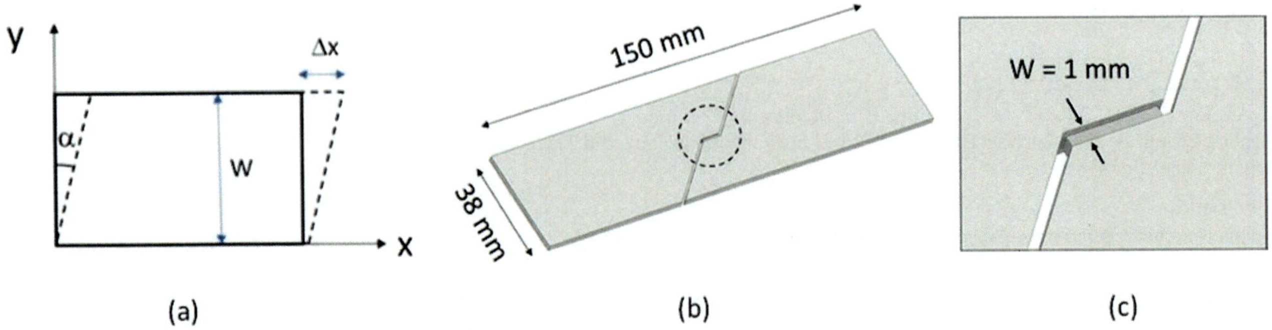

Fig. 19.1 (**a**) Definition of simple shear strain, $\gamma = \tan \alpha = (\Delta x)/W$. (**b**) Solid model of ASTM B831 specimen [9, 10], the total thickness is 2 mm. (**c**) Close up of the reduced gauge section in (**b**), with dimensions 6.35 mm long × 1 mm wide × 1 mm thick

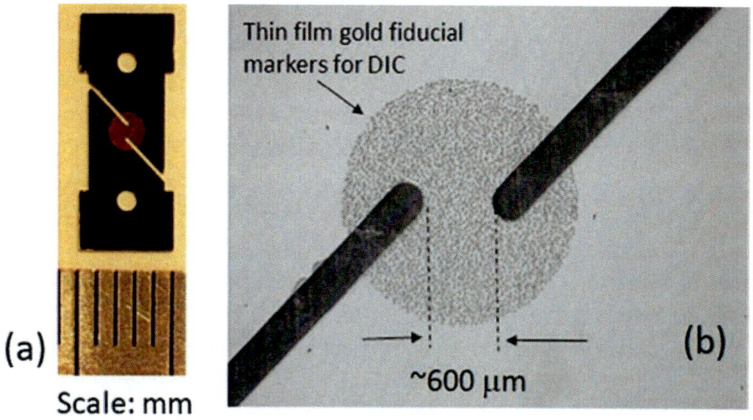

Fig. 19.2 (**a**) Photograph of a shear specimen beside a millimeter ruler. The specimens were 200 μm thick. (**b**) Optical micrograph of the gauge section

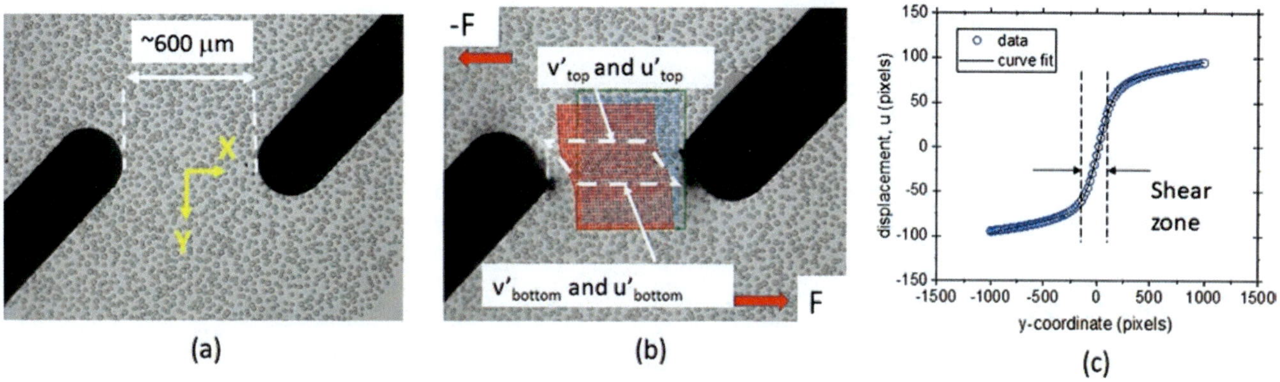

Fig. 19.3 Optical micrographs from a shear test, (**a**) at the start of the test (**b**) during simple shear deformation with the DIC grid overlayed, where *u* is the displacement in *x*, *v* is the displacement in *y*, and *u'* and *v*' are these displacements after a correction for rigid-body rotation has been applied. The trapezoid (dotted lines) was drawn manually to represent the shear zone and is meant to guide the eye only. (**c**) Typical displacement in the *x*-direction (*u*) vs. *y*-position, for one image

of uniform thickness to reduce the LIGA fabrication complexity. This design change made the measurement of the shear strain more challenging, as described later in this section. A typical specimen is shown in Fig. 19.2. As shown in Fig. 19.2b, an array of 25 um-diameter and 1 um-thick Au dots was deposited and patterned on the surface of the gauge section to create a speckle pattern to facilitate the use of digital image correlation (DIC) for strain measurement [7].

The specimens were pulled apart at strain rate 0.001/s in a commercial table-top miniature tension stage mounted under an optical microscope. Force was measured with a 334 N (75-lb) load cell, and optical micrographs of the gauge sections were captured by a CCD camera. For the higher rate tests, a larger commercial bench top load frame with a 5 kN load cell was used, with a high-speed camera for acquiring images. More details of the test apparatus are in [7]. The specimen thicknesses were measured, before testing, with a digital micrometer at the grip sections. The gauge width, designed at 600 μm, was measured from optical micrographs. The engineering shear stress was calculated as the force divided by the gauge section's initial cross-sectional area. Figures 19.3a, b show the gauge section of a specimen before and during a shear test. DIC, using a custom program, was used to measure the displacements across the gauge section. The forces and images were analyzed off line to construct the engineering shear stress vs. shear strain curves.

The ASTM B831 standard [9] defines the calculation of the simple shear strain from measurable quantities (displacements) using Eq. 19.1. However, since our specimens were of uniform thickness, the gauge width W in Eq. 19.1 and Fig. 19.1a was neither physically defined nor constrained as in the ASTM design (compare Fig. 19.3a with Fig. 19.1c) but instead would evolve during the course of a test. Determining the shear strain in our specimens was thus considerably more involved than in the ASTM standard. Our approach was to plot the DIC displacements (u) in the direction of shear, with respect to the transverse position, y, averaged across the DIC grid shown in Fig. 19.3b. From the resulting displacement plot (Fig. 19.3c), the slope of the linear section of the "S" shaped curve is the simple shear strain, γ. As the shear test proceeds, the "S" shaped curve evolves in size and slope. To automate the extraction of γ from all the images, a custom program was written to perform an empirical fit to the displacements using a hyperbolic tangent function:

$$u_{predicted} = y_0 + m(x - x_0) + Ctanh[D(x - e)] \tag{19.2}$$

where $u_{predicted}$ is the predicted value of the displacement, x is the independent variable which in this case is the y-coordinate in the DIC images, x_o is the starting y-coordinate at the beginning of the data set, and y_o, m, C, D, and e are constants found by nonlinear regression. The shear strain is then extracted by the program as:

$$\gamma = CD \tag{19.3}$$

In addition to the shear specimens, we also designed, fabricated and tested tensile specimens with gauge section dimensions of 2 mm long × 200 μm wide × 200 μm thick. The tensile specimens were of the same materials as the shear specimens but were fabricated in a separate fabrication run. The same testing apparatus was used as for the shear tests, and the same DIC program was used to measure the engineering strain from the displacements at the ends of the gauge sections. Details of the tensile tests are reported elsewhere [7, 8].

19.3 Results and Discussion

Figure 19.4 shows representative engineering shear stress versus simple shear strain curves, to fracture, for the two LIGA alloys, at strain rates of 0.001/s and 30/s. The Ni-10%Fe shows much higher shear strength while the Ni-10%Co alloy exhibits much higher ductility. The measurement uncertainty in the shear stress is under 2%. The measurement uncertainty in the shear strain for the Ni-Fe is within ~7%. The shear strain values for the Ni-Co alloy (Fig. 19.4b) are within ~4% uncertainty until the maximum shear stress. After the maximum stress is reached, as the ductile Ni-10%Co continues to be pulled apart the gauge section undergoes extensive plastic deformation, such that the deformation profile starts to deviate from the simple shear defined in Fig. 19.1, this deviation reaching a maximum of 35–40% at fracture. Nevertheless, the curves for the Ni-10%Co alloy are plotted for all strain values until fracture, for comparison with the Ni-10%Fe.

Figure 19.5 shows the fracture surfaces of a shear specimen of each material, tested at the lower strain rate. The Ni-Fe alloy (Fig. 19.5a) fails in a brittle manner, while the Ni-Co alloy (Fig. 19.5b) fails in a ductile manner. The fracture surfaces of *tensile* specimens of the same materials show similar characteristics [11]. The fracture mechanisms of the two alloys are discussed in detail in Ref [11].

Table 19.1 summarizes the shear and tensile properties of the two LIGA alloys, at two strain rates: a quasi-static strain rate of 0.001/s, and an intermediate strain rate of 30/s for the shear specimens, and 1/s for the tensile specimens. The tensile properties are discussed in more detail in [7, 8]. All nominal values in Table 19.1 are averaged over 6 shear specimens per material/rate combination, and 3 tensile specimens per material/rate combination. The uncertainties reported are the measurement uncertainties, calculated from error analysis, with the exception of the tensile specimens' true strain to fracture where the uncertainty is the standard deviation. Both materials exhibit an increase in the shear strength, tensile strength, and true strain to fracture, with increasing strain rate. The nominal values of the Young's modulus and shear modulus appear to

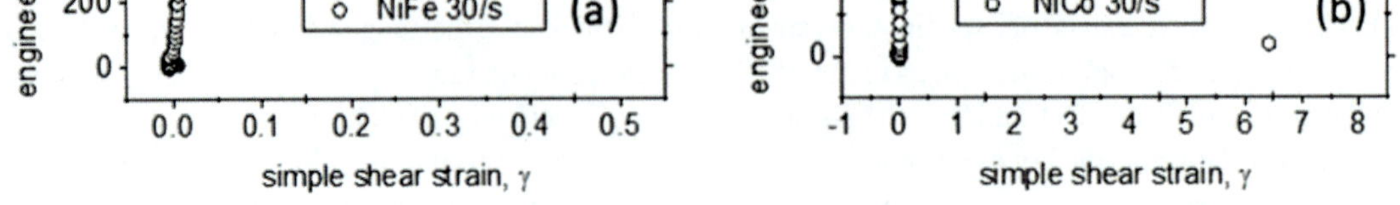

Fig. 19.4 Shear stress vs shear strain curves, to fracture, for (**a**) nanocrystalline Ni-10%Fe, (**b**) micrograined Ni-10%Co. For (**b**), the shear strains beyond the maximum shear stress are not accurate due to up to 40% deviation from simple shear in the specimen during the test. The curves are nonetheless plotted to failure for comparison to (**a**)

Fig. 19.5 Scanning electron micrographs of the fracture surfaces of shear specimens of (**a**) Ni-10%Fe and (**b**) Ni-10%Co

Table 19.1 Shear and tensile properties of two LIGA Ni alloys at two strain rates

Property	Quasi-static rate[a]	Quasi-static rate[a]	Intermediate rate[b]	Intermediate rate[b]
	Ni-10%Fe	Ni-10%Co	Ni-10%Fe	Ni-10%Co
Shear modulus (GPa)	48 ± 7%	33 ± 9%	40 ± 11%	37 ± 14%
Shear strength, τ_{max} (MPa)	1158 ± 2%	408 ± 1%	1205 ± 2%	455 ± 2%
Yield strength in shear (MPa)	615 ± 7%	173 ± 9%	720 ± 11%	197 ± 14%
Shear strain at fracture	0.42 ± 5%	N/A	0.54 ± 9%	N/A
Shear strain at τ_{max}	0.31 ± 7%	2.0 ± 4%	0.14 ± 7%	2.0 ± 2%
Young's modulus (GPa)	147 ± 12%	143 ± 26%	123 ± 81%	99 ± 75%
Tensile strength (MPa)	1949 ± 2%	604 ± 3%	2034 ± 2%	657 ± 3%
Yield strength (MPa)	1394 ± 12%	411 ± 26%	1609 ± 34%	432 ± 75%
True strain at fracture[*]	0.83 ± 0.11%	1.82 ± 0.31%	1.15 ± 0.03%	2.15 ± 0.22%

Values are averaged over 6 shear specimens per material/rate combination, and 3 tensile specimens per material/rate combination. The uncertainties are the measurement errors calculated from error analyses. The tensile specimens had gauge dimensions of 200 μm × 200 μm × 2 mm length
[*]The uncertainty in the true strain to fracture is the standard deviation. [a]0.001 /s. [b]1/s for tensile and 30/s for shear

exhibit rate effects but are inconclusive due to the large associated measurement uncertainties. It has been established in the conventional macroscale mechanical testing community that the Young's modulus is the least reliable property obtained from a tensile test [12] and is best measured using resonance or acoustic methods. The experimental factors that lead to large uncertainty in the Young's modulus are exacerbated for meso- and microscale tensile tests. Similar factors apply to the measurement of the shear modulus. A more detailed discussion of the measurement uncertainties associated with the tensile properties in Table 19.1, will be presented in a future paper.

Figure 19.6 shows the equivalent plastic strain at fracture for the tensile and shear specimens of the nanocrystalline Ni-10%Fe alloy, as a function of their respective stress triaxialities. The triaxiality χ is given by:

$$\chi = \frac{\sigma_H}{\sigma_{VM}} \tag{19.4}$$

where σ_H is the hydrostatic stress and σ_{VM} is the Von Mises stress. The effective strain at fracture for the shear specimens was estimated using the measured shear strain at fracture and the Von Mises yield criterion. The equivalent plastic strain for the Ni-10%Fe tensile specimens was obtained by use of finite element analysis [13]. The hydrostatic and Von Mises stresses were calculated from the principal stresses, assuming uniform and ideal stress states (i.e., pure tension for the tensile specimens and pure shear for the shear specimens). More accurate values would be obtained using finite element analysis. The tensile and shear specimens thus provide two points for constructing the fracture locus of the Ni-10%Fe alloy, as shown in Fig. 19.6, supporting the development of failure prediction models for LIGA MEMS devices. Note that Fig. 19.6 violates conventional failure prediction models for ductile metals in which the fracture strain typically increases with decreasing triaxiality. The trend in Fig. 19.6 might be due to the influence of the Lode angle parameter, as shown in recent studies at the macroscale [14].

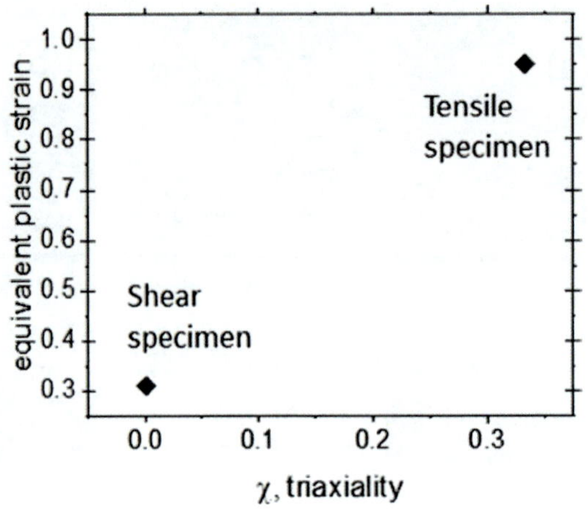

Fig. 19.6 Two points on the fracture locus for 200 μm-thick nanocrystalline Ni-10%Fe

19.4 Conclusion

We reported the design, fabrication and tensile- and shear testing of LIGA nanocrystalline Ni-10%Fe and micrograined Ni-10%Co, at two strain rates. The shear specimens were 200 μm thick and had gauge widths of 600 μm. The tensile specimens had gauge dimensions of 2 mm × 200 μm × 200 μm. Engineering shear stress vs simple shear strain curves were obtained for both alloys, at strain rates of 0.001/s and 30/s. The higher strain rate resulted in increased shear strength, tensile strength, and true strain to fracture, for both materials. We tabulated the shear properties and tensile properties along with their measurement uncertainties, and constructed part of the fracture locus for the nanocrystalline Ni-10%Fe. It is anticipated that this data may support the development of constitutive models for structural failure predictions of LIGA MEMS devices.

Acknowledgements Specific commercial equipment, instruments, and materials that are identified in this report are listed in order to adequately describe the experimental procedure and are not intended to imply endorsement or recommendation by the National Institute of Standards and Technology (NIST). This project was funded by the U.S. Army Combat Capabilities Development Command Armaments Center (CCDC AC) and the Joint Fuze Technology Program (JFTP). The JFTP project number was 14-G-006, Micro Scale Materials and Energetic Effects Characterization.

References

1. Becker, E.W., Ehrfeld, W., Hagmann, P., Maner, A., Munchmeyer, D.: Fabrication of microstructures with high aspect ratios and great structural heights by synchrotron radiation lithography, galvanoformung, and plastic moulding (LIGA process). Microelec. Eng. **4**, 35–36 (1986)
2. Ehrfeld, W., Bley, P., Gotz, F., Hagmann, P., Maner, A., Mohr, J., Moser, H.O., Munchmeyer, D., Schelb, W., Schmidt, D., Becker, E.W.: Fabrication of microstructures using the LIGA process. Proc. 1987 IEEE Micro Robots and Teleoperators Workshop, Hyannis, MA, pp. 11/1–11/11 (1987)
3. Schiltges, G., Gsell, D., Dual, J.: Torsional tests on microstructures: two methods to determine shear-moduli. Microsys. Tech. **5**, 22–29 (1998)
4. Schiltges, G., Dual, J.: Failure behavior of microstructures under torsional loads. J. Mech. Phys. Solids. **49**, 1021–1038 (2001)
5. Saif, M.T.A., MacDonald, N.C.: Measurement of forces and spring constants of microinstruments. Rev. Sci. Inst. **69**, 1410–1422 (1998)
6. Pantano, M.F., Espinosa, H.D., Pagnotta, L.: Mechanical characterization of materials at small length scales. J. Mech. Sci. Tech. **26**, 545–561 (2012)
7. Liew, L.A., Read, D.T., White, R.M., Barbosa, N.: U.S. Army ARDEC Joint Fuze Technology Program (JFTP) Task 2 report: quasi-static tensile tests of microfabricated electrodeposited (LIGA) Ni alloys. NIST interagency report 8182, (2018).https://doi.org/10.6028/NIST.IR.8182
8. L.A. Liew, D.T. Read, M.L. Martin, F.W. DelRio, P.E. Bradley, N. Barbosa, T.R. Christenson, J.T. Geaney. Elastic-plastic properties of mesoscale electrodeposited LIGA nickel alloys films: microscopy and mechanics. in preparation, J. Mech. Microeng. **31**, 015002, (2020)
9. ASTM Committee B07: Standard test method for shear testing of thin aluminum alloy products (ASTM B831–14). ASTM International, West Conshohocken, PA (2014)
10. Kang, J., Wilkinson, D.S., Wu, P.D., Bruhis, M., Jain, M., Embury, J.D., Mishra, R.K.: Constitutive behavior of AA5754 sheet materials at large strains. ASME J. Eng. Mat. Tech. **130**, 031004-1-5 (2008)
11. Martin, M.L., Liew, L.A., Read, D.T., Christenson, T., DelRio, F., Geaney, J.: Dominant factors for fracture at the microscale in electrodeposited nickel alloys. Sens. Actuators A: Phys. 2020;314:112239
12. Lord, J.D., Morrell, R.: Measurement Good Practice Guide No. 98: Elastic Modulus Measurement (National Physical Laboratory, UK). (2006)

13. Read, D.T., Liew, L.A., White, R.M., Barbosa, N., Geaney, J.: Measuring flow curve and failure conditions for a MEMS-scale electrodeposited nickel alloy. Mater. Res. Exp. **6**, 0950a6 (2019)
14. Bao, Y., Wierzbicki, T.: On fracture locus in the equivalent strain and stress triaxiality space. Int. J. Mech. Sci. **46**, 81–98 (2004)